Public-Sector Aviation: Graduate Research Award Papers 2018

Transportation Research Record: Journal of the Transportation Research Board Volume 2672, Issue 29

December 2018

ISBN: 978-1-5108-9205-7

Printed from e-media with permission by:

Curran Associates, Inc.
57 Morehouse Lane
Red Hook, NY 12571

Some format issues inherent in the e-media version may also appear in this print version.

Published by Sage Publications USA

ISBN (Print) 978-1-5108-9205-7
ISBN 2018 Printed Set (All Issues) 978-1-5108-7735-1

Printed with permission by Curran Associates, Inc. (2019)

Additional copies of this publication are available from:

Curran Associates, Inc.
57 Morehouse Lane
Red Hook, NY 12571 USA
Phone: 845-758-0400
Fax: 845-758-2633
Email: curran@proceedings.com
Web: www.proceedings.com

TRR
JOURNAL OF THE TRANSPORTATION RESEARCH BOARD

Article

Transportation Research Record
2018, Vol. 2672(29) 1–11

DOI: 10.1177/0361198118780876
journals.sagepub.com/home/trr
SAGE

Data-Driven Modeling of Aircraft Engine Fuel Burn in Climb Out and Approach

Yashovardhan S. Chati[1] and Hamsa Balakrishnan[1]

Abstract

Fuel burn is a key driver of aircraft performance, and contributes to airline costs and emissions. Low-altitude fuel burn and emissions, such as those that occur during climb out and approach, have a significant impact on the environment in the vicinity of airports. This paper proposes a new methodology to statistically model fuel burn in the climb out and approach phases using the trajectory of an aircraft. The model features are chosen by leveraging a physical understanding of aircraft and engine dynamics. Model development is conducted through the use of Gaussian Process Regression on a limited Flight Data Recorder archive, which also provides ground truth estimates of the fuel flow rate and total fuel burn. The result is a class of models that provide predictive distributions of the fuel burn corresponding to a given aircraft trajectory, thereby also quantifying the uncertainty in the predictions. The performance of the proposed models is compared with other frequently used Aircraft Performance Models. The statistical models are found to reduce the error in the estimated total fuel burn by more than 73% in climb out and by 59% in approach.

The amount of fuel consumed, as well as the rate of consumption, are key aspects of an aircraft's performance. Fuel burn also constitutes a major component of the Direct Operating Costs of an airline, and results in emissions that have environmental and health impacts, especially in the vicinity of airports.

It is estimated that about 10% of all aircraft emissions, and 30% of aircraft carbon monoxide emissions, occur close to the ground (*1*). Climb out can contribute to as much as 6% of the total fuel consumed in a flight, while approach can contribute to as much as 4% of the total fuel consumed. It is therefore important to estimate the fuel burn (and subsequently, the emissions) in the climb out and approach phases of flight. While fuel flow profiles are typically monitored by sensors on board aircraft, they are not disseminated to entities outside the airline, including environmental inventories. By contrast, aircraft trajectory information (i.e., its physical location as a function of time) is available from surveillance systems. Motivated by these observations, this paper investigates the following problem: *Given the trajectory of a flight, can one infer its fuel burn profiles in the climb out and approach phases?*

The ability to translate aircraft trajectory information into fuel burn estimates forms an important component of fuel burn and emissions inventories. Such tools can be used by aviation researchers to assess the impact that the introduction of new takeoff and landing procedures will have on the environment around airports. Projection studies make use of aircraft performance models to forecast the emissions impacts of future demand growth (*2*).

Engine performance simulators (e.g., GasTurb (*3*)) typically use physics-based approaches to model fuel burn. Such approaches are helpful in design studies, since it is possible to control the underlying simulation variables needed as inputs to the model. However, they require a knowledge of engine operating points, and parameters such as the temperature and pressure ratios at any time, which are not readily available for operational flights. Previous efforts to develop data-driven models of fuel burn have typically used simulated or nonoperational data from flight manuals, ground tests, simulation, and performance calculators to model the fuel flow rate (4, 5). Such methods do not always capture the performance of real-world aircraft operations. Prior work has also shown that the International Civil Aviation Organization (ICAO) Aircraft Engine Emissions Databank (ICAO Databank) (*6*) can overestimate operational fuel burn and emissions, and that the total fuel burn in the Landing and Takeoff (LTO) cycle can be overestimated by up to 40% (*7*). Finally, there

[1]Massachusetts Institute of Technology, Cambridge, MA

Corresponding Author:
Address correspondence to Yashovardhan S. Chati: yschati@mit.edu

has been recent work categorizing different international efforts on fuel burn modeling, and describing the associated challenges (*8*).

An aircraft in flight is subject to random disturbances, such as manufacturing tolerances, component deterioration, air turbulence, and so forth. As a result, the aircraft is better modeled as a statistical system than as a deterministic one (*9*). A statistical model also enables the quantification of uncertainty in the fuel burn. These uncertainty estimates reflect the effect of random disturbances and modeling assumptions, and help quantify the variability seen in real operations. The quantification of the uncertainty in nominal performance can help identify and rectify anomalous behavior, thereby saving significant maintenance costs (*10*).

Contributions of This Paper

The primary contribution of this paper is the development of a methodology for the data-driven statistical modeling of aircraft fuel burn in climb out and approach. The fuel burn in these phases of flight has a particularly significant environmental impact on airports and their surroundings. The modeled variables are the instantaneous fuel flow rate at any time, and the total mass of fuel consumed (in climb out or approach). Both these quantities are important from an emissions perspective: some emissions (e.g., nitrogen oxides, hydrocarbons) depend on the instantaneous fuel flow rate, whereas others (e.g., carbon dioxide, water vapor) depend on the total fuel burn.

The modeling approach adopted in this paper attempts to use data to identify the mapping between the input and output variables. To identify such a mapping, a training dataset is required that contains the input (trajectory) variables for a set of flights, as well as the corresponding realized fuel flow profiles. In other words, a supervised learning approach (*11*) is adopted. The model will then, given a new set of trajectory variables, predict the corresponding fuel flow profile. The resulting predictive distribution will also reflect the variability seen in the training data and model uncertainty.

The proposed models leverage a physical understanding of aircraft and engine performance to identify the important features for fuel burn estimation. Reasonable assumptions are made to restrict the feature space to trajectory variables. The models are based on a statistical technique known as Gaussian Process Regression (*12*). The fuel burn estimates from the developed models are compared with those given by widely-used alternatives such as the Base of Aircraft Data (BADA), the Senzig-Fleming-Iovinelli (SFI), and the Boeing Fuel Flow Method 2 – corrected ICAO Databank (ICAO-BFFM2) models. The statistical models proposed in this paper are found to reduce errors in the estimation of the mean fuel flow rate by more than 62% in climb out and 34% in approach, averaged over the different aircraft types. Furthermore, these improved fuel flow rate estimates reduce the error of the estimated total fuel burn by more than 73% in climb out, and 59% in approach (averaged over the different aircraft types). It should be noted that, in this paper, climb out and approach are taken to be the parts of ascent and descent, respectively, taking place below 3000' Above Field Elevation.

Dataset

A supervised learning approach requires a training dataset that contains the input trajectory variables, as well as the realized fuel flow rate profiles. A dataset obtained from the Flight Data Recorders (FDRs) of operational flights of a European airline is used. The FDR is usually an accurate data source, since it records parameters during flight. The sampling frequency of the FDR data is 1 Hz during the takeoff, initial ascent, late descent and landing phases, and as low as 1/300 Hz in portions of the cruise phase. Eight different aircraft/engine types are represented in the dataset (Table 1): six Airbus aircraft types (whose engine types are derivable from their series and mark designators), Boeing B767-300 (run by GE CF6-80C2B7F engines), and B777-300ER (run by GE90-115B1 engines).

Model Features

The features of a model refer to the variables (and their transformations) that are included. Instead of a purely data-driven feature identification procedure, this paper leverages a physical understanding of aircraft dynamics. Four principal forces act on an aircraft in flight, namely, the thrust generated by the engines which propels the aircraft forward, the lift generated (primarily by the wings) which keeps it in the air, the drag or the resistance offered by the air, and the weight of the aircraft. Considering the aircraft as a point mass and using a simplified model of aircraft dynamics, the aircraft net thrust from all the engines (F_n) is given as

$$F_n = qSC_{D_0} + \frac{C_{D_2}m^2g^2}{qS} - \frac{C_{D_2}m^2g^2\dot{h}^2}{qSV^2} + mg\frac{\dot{h}}{V} + m\frac{dV}{dt}. \tag{1}$$

Here, S is the reference wing area (a constant for a particular aircraft type), C_{D_0} and C_{D_2} are aircraft drag coefficients, m is the aircraft gross mass, g is the acceleration due to gravity, $\dot{h}$ is the vertical speed, V is the true airspeed, and t

is the time. q is the aircraft dynamic pressure and equals $\frac{1}{2}\rho_\infty V^2$, where ρ_∞ is the ambient air density.

The average fuel flow rate per engine ($\dot{m}_f$) can be related to the net thrust via the Thrust Specific Fuel Consumption (TSFC) as

$$\dot{m}_f = \frac{\text{TSFC} \times F_n}{N_{\text{eng}}}, \quad (2)$$

where N_{eng} is the number of engines. For a particular engine type, the corrected TSFC can be assumed to be a function of the aircraft Mach number, the corrected engine parameters (such as corrected net thrust for the engine), and the engine component efficiencies (*13, 14*), and is given by

$$\frac{\text{TSFC}}{\sqrt{\theta_\infty}} = f_{\text{TSFC}}(M_\infty, \frac{F_{n,pe}}{\delta_\infty}, \nu_{\text{eng}}). \quad (3)$$

Here, $M_\infty = \frac{V}{\sqrt{\gamma R T_\infty}}$ is the aircraft Mach number, δ_∞ is the ambient pressure (P_∞) divided by the International Standard Atmosphere (ISA) sea level static pressure (101,325 Pa), θ_∞ is the ambient temperature (T_∞) divided by the ISA sea level static temperature (288.15 K), $F_{n,\,pe}$ is the averaged net thrust per engine, and ν_{eng} represents the engine component efficiencies. γ is the adiabatic constant for air and R is the gas constant for air. Equations 1–3 reveal the following functional dependency for the averaged per engine fuel flow rate for a given aircraft/engine type (neglecting constants for the particular aircraft/engine type):

$$\dot{m}_f|_{\text{aircraft/engine type}} = f_{\dot{m}_f}(qS, m, \frac{\dot{h}}{V}, V, \frac{dV}{dt}, \sqrt{T_\infty}, C_{D_0}, C_{D_2}, \nu_{\text{eng}}) \quad (4)$$

The aim of this study is to develop aircraft type specific models which can map the aircraft trajectory to fuel flow rate profiles. Trajectory data typically comprise the aircraft latitude, longitude, altitude, ground speed, climb rate, and course as functions of time. These models can then be used to estimate the fuel flow rate for a flight, given its trajectory from any surveillance system. The following simplifying assumptions are made to restrict the feature set, to the extent possible, to features which are extractable from trajectory data:

- Since high-fidelity weather data may not be available, the ambient air density, pressure, and temperature are assumed to be a function of the altitude, according to the ISA assumptions. Consequently, only the density is retained in the model to represent ambient weather conditions.
- Aircraft performance depends on the true airspeed (V), which is the aircraft velocity with respect to the surrounding air. Due to the effect of wind, this speed differs from the ground speed, namely, the speed relative to the ground. The trajectory of an aircraft, as observed by a surveillance system, only provides information on the ground speed. The calculation of the true airspeed requires additional wind data of the same spatial and temporal resolution as the trajectory information which is often unavailable. As the value of the true airspeed cannot be derived from the trajectory with the same fidelity as the other variables, it is not included as an input feature. Instead, the ground speed (V_{GS}), which can be derived from trajectory data, is used as a feature. It is expected that the neglect of the effects of wind will be reflected in larger variability, and consequently, larger prediction errors as well as prediction intervals.
- The rate of change of the true airspeed ($\frac{dV}{dt}$) is also not observed in surveillance data. The numerical derivative of the ground speed ($\frac{\Delta V_{\text{GS}}}{dt}$)is therefore used as a feature. The values of the numerical derivative are smoothed through a low pass filter before using them for analysis to remove noise arising from numerical differentiation.
- The BADA assumptions on the drag coefficients are assumed to hold. In climb out, the drag coefficients are considered to be aircraft-type specific constants. In approach, the drag coefficients are assumed to have discrete levels depending on the altitude of the aircraft with respect to the mean sea level elevation of the arrival airport (h_{ATD}), the aircraft speed, and the aircraft gross mass. This dependence results from the different configurations the aircraft is in as it comes in to land (due to deployment of flaps, slats, landing gear) which change the aircraft drag coefficients.
- The engine component efficiencies depend on factors like the engine operating point, the engine age, and the level of maintenance. Because trajectory data alone do not give information on the engine component efficiencies or factors influencing the efficiencies, ν_{eng} is not included as a predictor variable.

As a result of the above assumptions, the functional relation in Equation 4 can be simplified to the following relation for a given aircraft/engine type:

$$\dot{m}_f|_{\text{aircraft/engine type}} \approx \begin{cases} \dot{m}_{f(\text{co})}(q_{\text{GS}}S, m, \frac{\dot{h}}{V_{\text{GS}}}, V_{\text{GS}}, \frac{\Delta V_{\text{GS}}}{\Delta t}), & \text{in climb out, and} \\ \dot{m}_{f(\text{ap})}(q_{\text{GS}}S, m, \frac{\dot{h}}{V_{\text{GS}}}, V_{\text{GS}}, \frac{\Delta V_{\text{GS}}}{\Delta t}, h_{\text{ATD}}), & \text{in approach.} \end{cases} \quad (5)$$

Here, the subscripts (co) and (ap) refer to the climb out and approach phases, respectively. q_{GS} is the dynamic pressure based on the ground speed, given by $\frac{1}{2}\rho_\infty V_{GS}^2$. The predictor/input variables are therefore the dynamic pressure multiplied by the reference wing area ($q_{GS}S$), the aircraft mass (m), the ratio of the vertical speed to the ground speed ($\frac{\dot{h}}{V_{GS}}$), the ground speed (V_{GS}), and the rate of change of the ground speed ($\frac{\Delta V_{GS}}{dt}$). In approach, the aircraft altitude above the arrival airport elevation (h_{ATD}) is also taken as an additional predictor variable. The predicted/output variable is the aircraft per engine fuel flow rate ($\dot{m}_f$), averaged over all the engines. All the variable values are taken in SI units. It is important to note that except for the aircraft mass, all the predictor variables can be extracted from trajectory surveillance data. The cumulative effect of random disturbances, unmodeled factors, and different assumptions will be reflected in the fuel flow rate prediction intervals.

Statistical Models

Since all the variables chosen for the model are continuous, metric variables, mapping the aircraft trajectory to its fuel flow rate profile using a subset of labeled operational data comprises a supervised learning regression problem (*11*). In regression, the output of the ith observation (y_i) is assumed to equal a function of the input features ($\mathbf{x_i}$) plus some noise (ε_i),

$$y_i = f(\mathbf{x_i}) + \varepsilon_i. \tag{6}$$

The goal is to estimate the underlying regression function ($f(\mathbf{x})$) using a set of given inputs and their corresponding given outputs. Once the regression function is estimated, it can be used to estimate the unknown output corresponding to a new input.

For the fuel flow rate modeling problem being investigated in this paper, a regression method is desired that is both nonparametric and probabilistic. Nonparametric methods adapt the regression function to the system complexity represented by the data, without requiring a choice of the form of the regression function a priori. Thus, nonparametric methods are useful in problems with no prior knowledge of the exact functional form of the feature vectors (which may be nonlinear). Probabilistic algorithms assume the noise terms to follow an underlying distribution. They can thus, give the complete predictive distribution of the output at a new input vector. A nonparametric, probabilistic regression method found to be useful in this paper is Gaussian Process Regression (GPR). GPR has found application in diverse areas, ranging from biomedical applications and health care (*15*) to music (*16*).

Gaussian Process Regression (GPR)

In GPR, the regression function ($f(\mathbf{x})$) is assumed to be drawn from a Gaussian Process (GP) with zero mean function and covariance/kernel function $k(\mathbf{x}, \mathbf{x}')$.

$$f(\mathbf{x}) \sim \mathcal{GP}(0, k(\mathbf{x}, \mathbf{x}')) \tag{7}$$

A function is said to follow a Gaussian Process when any finite set of function values follows a joint Gaussian distribution (*12*). The noise (ε_i) is also assumed to follow a Gaussian distribution. The function $k(\mathbf{x}, \mathbf{x}')$ is called the kernel function, and represents the covariance between the regression function values $f(\mathbf{x})$ and $f(\mathbf{x}')$ at two inputs $\mathbf{x}$ and $\mathbf{x}'$. The flexibility in GPR arises from the rich variety of possible kernel functions. Two kernel functions which have been found to be particularly useful in this study are the Dot Product Squared Exponential (DPSE) and the Dot Product Exponential (DPE) kernels. Kernel functions depend on *hyperparameters* which govern behavior such as their characteristic length-scales. Model building using GPR involves estimating these hyperparameters and thereby, the form of the kernel function. Details on kernel functions and hyperparameters can be found in prior work (*17*). Once the hyperparameters are estimated, GPR can be used to give the predictive distribution of the output at a new input vector. The assumptions of Gaussianity (Gaussian Process and Gaussian noise) lead to a Gaussian predictive distribution, resulting in mathematical tractability. More details of GPR can be found in (*12*).

Model Building

The FDR dataset for each aircraft type (for the climb out and approach phases) is divided into three parts, with flights randomly selected to belong to each subset. The training set (65% of flights) is used to build the model, the validation set (15% of flights) is used to select the 'best' model from a set of candidate models, and the test set (20% of flights) is used to evaluate the predictive performance of the model that is finally chosen.

Equation 5 shows that the aircraft mass is an input/predictor variable for estimating the fuel flow rate. Instead of using the instantaneous aircraft mass as a predictor variable, the mass of the aircraft at takeoff (i.e., the takeoff weight (TOW)) is used as a predictor variable. Aircraft mass is typically not known for a particular flight. The mass of the aircraft changes during flight due to the consumption of fuel. By using the TOW as the predictor variable, the set of unknown predictor variables is restricted to the mass of the aircraft at a single instant (the takeoff instant) rather than the mass of the aircraft at each instant in flight. Moreover, using the TOW as a predictor variable enables simultaneous prediction of the fuel flow rate at all instants in climb out or approach (i.e., batch prediction) leading to faster predictions. Thus, this algorithm of model building is named the 'batch prediction' algorithm. In this paper, it is assumed that the TOW is known (the value recorded in the FDR data is used as an input feature). However, the TOW of a particular flight is not known in practice. In recent work by us, it is demonstrated how GPR can be used to estimate distributions of the TOW from the takeoff roll trajectory variables (*18*).

Using the true values of all input and output variables in the training dataset, GPR models with different kernel functions are trained for each aircraft type, for the climb out and approach phases. All the variables are standardized before training, that is, they are shifted by the sample mean and then scaled by the sample standard deviation of the respective variables in the training datasets. The GPR modeling is done using the MATLAB®-based *GPstuff* toolbox (*19, 20*).

Algorithms using the instantaneous aircraft mass as a predictor variable have also been considered in previous work (*17*). These require a one-step tandem prediction of both the aircraft mass and the fuel flow rate at each time instant. Such algorithms are found to be computationally slower, and yet not give significantly better predictive performance than the batch prediction algorithm, on unseen data. Thus, they are not considered further in this paper and only the batch prediction algorithm is used.

Model Evaluation

The main objective of this paper is the development of a model that predicts the fuel flow rate profile for a flight, given its trajectory data. Given the input variables for a flight, the GPR models produce a predictive distribution of the fuel flow rate. The accuracy of both the point and the interval estimates of the predicted fuel flow rate are evaluated using the following metrics:

- **Mean Absolute Error (MAE)**: This is the mean of the absolute values of the relative prediction errors on the unseen prediction dataset, given by

$$MAE = \frac{1}{n^*}\sum_{i=1}^{n^*}\left|\frac{\hat{\dot{m}}_{f\,i} - \dot{m}_{f\,i,\text{true}}}{\dot{m}_{f\,i,\text{true}}}\right|. \tag{8}$$

Here, n^* is the number of observations in the unseen prediction set, $\dot{m}_{fi,\text{true}}$ is the actual fuel flow rate (ground truth), and $\widehat{\dot{m}}_{fi}$ is the model mean prediction of the fuel flow rate, for the ith observation in the prediction set. The model mean prediction is the mean of the Gaussian fuel flow rate predictive distribution. The MAE therefore reflects the accuracy of the mean prediction (point estimate). A model with a low MAE is desired.

- **Prediction Coverage (PC)**: This is the fraction of observations in the unseen prediction set for which the ground truth values of the fuel flow rates fall within the 95% prediction intervals of the fuel flow rate predictive distributions. The 95% prediction intervals are given by the 95% Highest Density Intervals (HDIs) of the fuel flow rate predictive distributions. The PC indicates how well the prediction intervals capture the variability of the fuel flow rates. A PC value close to 95% increases belief that the model has been properly specified and formulated.
- **Normalized Length of Prediction Interval (NLPI)**: This is the mean of the length of the 95% prediction intervals expressed as a fraction of the corresponding mean fuel flow rate predictions. The NLPI indicates the extent of relative uncertainty present in the predicted fuel flow rate.

For each aircraft type in each phase of flight, GPR models using different kernel functions are trained, and applied to the validation dataset. The evaluation metrics for model predictive performance using different kernels are compared using statistical multiple comparison tests, and the kernel giving the best statistically significant predictive performance (at a 5% significance level) is chosen.

The GPR model with the chosen kernel function is finally evaluated for its predictive performance on the unseen test dataset. The Gaussian predictive distributions given by the GPR models on new inputs in the test data are used to derive the mean predicted fuel flow rates and the 95% prediction intervals for the fuel flow rate. It is worth noting that although GPR models are slow to train, they give fast predictions on new inputs (in the order of a few seconds). In other words, the time taken to predict the fuel flow rate profile for a new flight trajectory is small.

Existing Fuel Burn Models

The statistical models developed are also compared with current state-of-the-practice Aircraft Performance Models (APMs) that are used to estimate fuel burn in systems such as the FAA's Aviation Environmental Design Tool (AEDT). These include:

- **Base of Aircraft Data (BADA) model (*21*)**: Developed by EUROCONTROL, BADA is a total energy-based method used for aircraft performance modeling. It uses simplified equations to model aircraft performance, with different equation coefficients (maintained in the BADA database) for different aircraft types. This method needs aircraft net thrust values to estimate the fuel flow rate. In this paper, net thrust (F_n) values are determined using the BADA Family 3 (revision 3.13) thrust estimation equations (*21*). In this study, BADA 3.13 is used as just a benchmark model. BADA Family 4 models are not explored in this study and could potentially give diferent results than the BADA 3.13 models. The climb out net thrust ($F_{n_{co}}$) is given by the following set of equations:

$$F_{n_{co}} = (1 - C_{Tc,5}\Delta T_{\text{eff}})C_{Tc,1}(1 - \frac{h}{C_{Tc,2}} + C_{Tc,3}h^2)$$

$$\Delta T_{\text{eff}} = \Delta T - C_{Tc,4}$$

$$0 \leq \Delta T_{\text{eff}} \times C_{Tc,5} \leq 0.4$$

$$C_{Tc,5} \geq 0 \qquad (9-12)$$

Here, $C_{T\ c,1}$, $C_{T\ c,2}$, $C_{T\ c,3}$, $C_{T\ c,4}$, and $C_{T\ c,5}$ are thrust coefficients enumerated in the BADA database, and ΔT is the temperature deviation from the ISA. h is the aircraft altitude above mean sea level. The approach net thrust ($F_{n_{ap}}$) is given as:

$$F_{n_{ap}} = \begin{cases} C_{\text{Tdes, high}} \times F_{n_{co}}, & \text{if h>h}_{\text{de}}, \\ C_{\text{Tdes, low}} \times F_{n_{co}}, & \text{if h<h}_{\text{de}} \text{ \& aircraft is in cruise configuration,} \\ C_{\text{Tdes, app}} \times F_{n_{co}}, & \text{if h<h}_{\text{de}} \text{ \& aircraft is in approach configuration,} \\ C_{\text{Tdes, ld}} \times F_{n_{co}}, & \text{if h<h}_{\text{de}} \text{ \& aircraft is in landing configuration.} \end{cases} \qquad (13)$$

Here, $C_{\text{Tdes,high}}$, $C_{\text{Tdes,low}}$, $C_{\text{Tdes,app}}$, and $C_{\text{Tdes,ld}}$ are descent thrust coefficients enumerated in the BADA database. h_{de} is the transition altitude for calculation of descent phase thrust and is enumerated in the BADA database. The definitions of the different configurations are given in the BADA manual. These configurations are treated differently due to different aerodynamic drag values arising from the extension of slats, flaps, and landing gear. Once the net thrust is determined, the following equations are used to estimate the fuel flow rate per engine ($\dot{m}_{f\ \text{BADA}}$):

$$\dot{m}_{f\ \text{BADA}} = \begin{cases} \dfrac{\text{TSFC} \times F_n}{N_{\text{eng}}}, & \text{in climb out, and} \\ \dfrac{1}{N_{\text{eng}}} max(\text{TSFC} \times F_n,\ \dot{m}_{f\,\text{min}}), & \text{in approach.} \end{cases}$$

$$\text{TSFC} = C_{f1}(1 + \frac{V}{C_{f2}})$$

$$\dot{m}_{f\,\text{min}} = C_{f3}(1 - \frac{h}{C_{f4}}) \qquad (14)$$

Here, $\dot{m}_{f\,\text{min}}$ is the minimum aircraft fuel flow rate from all engines. C_{f1}, C_{f2}, C_{f3}, and C_{f4} are aircraft-type specific constants found in the BADA database. In the analysis done here, the values of the input variables to the BADA equations (such as, altitude, temperature, density, true airspeed, etc.) have been obtained from the FDR dataset.

- **Senzig-Fleming-Iovinelli (SFI) model (*4*)**: The SFI model is used in AEDT to model the fuel flow rate in the terminal region of ascent and descent (< 10000' AFE). The fuel flow rate per engine ($\dot{m}_{f\ \text{SFI}}$) is expressed as:

$$\dot{m}_{f\,\text{SFI}} = \begin{cases} (K_1 + K_2M_\infty + K_3h + K_4\frac{F_{n,pe}}{\delta_\infty})\sqrt{\theta_\infty}F_{n,pe}, & \text{in climb out, and} \\ (\alpha + \beta_1 M_\infty + \beta_2 \exp(-\frac{\beta_3\frac{F_{n,pe}}{\delta_\infty}}{F_{n_0}}))\sqrt{\theta_\infty}F_{n,pe}, & \text{in approach.} \end{cases} \qquad (15)$$

Here, $F_{n,pe}$ is the aircraft net thrust per engine, and F_{n_0} is the engine ISA sea level static thrust. K_1, K_2, K_3, K_4, α, β_1, β_2, and β_3 are aircraft-type specific model coefficients enumerated in the AEDT database. The per-engine net thrust ($F_{n,pe}$) values required as inputs to the SFI model are determined using the BADA thrust estimation equations, Equations 9–13.

- **The ICAO Databank with Boeing Fuel Flow Method 2 Correction (ICAO-BFFM2) model (*22*)**: The ICAO Databank tabulates the values of fuel flow rates in the LTO cycle (*6*). These values are obtained through ground-based uninstalled engine certification tests and the measurements are reported after being corrected to sea level static ISA reference conditions. The databank entry for each engine reports four values for the fuel flow rate, namely, at takeoff (100% thrust, which is seldom true as most takeoffs happen at derated thrust), climb out (85% thrust), approach (30% thrust), and ground idle/taxi (7% thrust, whereas in reality, many idling conditions occur at less than 7% thrust (*23*)). The Boeing Fuel Flow Method 2 (BFFM2) provides corrections to convert the ICAO Databank values to at-altitude conditions for an installed engine (*22*). The fuel flow rate is then given by the following equations:

$$\dot{m}_{f\text{ICAO-BFFM2}} = \begin{cases} 1.013\dot{m}_{f\text{ICAO, co}}\delta_\infty\theta_\infty^{-3.8}\exp(-0.2M_\infty^2), & \text{in climb out, and} \\ 1.020\dot{m}_{f\text{ICAO, ap}}\delta_\infty\theta_\infty^{-3.8}\exp(-0.2M_\infty^2), & \text{in approach.} \end{cases} \qquad (16)$$

Here, $\dot{m}_{f\text{ICAO-BFFM2}}$ is the ICAO-BFFM2 fuel flow rate per engine, and $\dot{m}_{f\text{ICAO}}$ is the reported value of the fuel flow rate for an uninstalled engine at reference conditions in the ICAO Databank. The subscripts "co" and "ap" refer to the reported ICAO Databank values in climb out

Table 1. Fuel Flow Rate: Predictive Performance of the GPR Models and Other Aircraft Performance Models (APMs) in Climb Out and Approach for Different Aircraft Types

			Existing APMs				GPR models	
			Mean absolute error (%)				PC (%)	NLPI (%)
Phase	Aircraft type	Number of flights	BADA	SFI	ICAO	GPR	GPR models	
Climb out	A319-112	130	4.7 (3.3)	8.2 (1.4)	13.0 (5.2)	2.6 (1.0)	95.6 (6.8)	13.9 (1.0)
	A320-214	169	7.3 (4.2)	9.7 (2.8)	8.6 (4.1)	3.5 (2.5)	94.7 (9.8)	20.7 (1.2)
	A321-111	117	11.8 (3.5)	14.6 (2.6)	13.1 (4.1)	4.3 (1.7)	91.7 (9.0)	21.7 (1.2)
	A330-202	84	11.1 (1.6)	10.2 (1.4)	17.3 (2.9)	4.1 (1.8)	91.8 (9.7)	18.3 (1.3)
	A330-243	100	10.3 (1.7)	–	5.9 (3.2)	3.1 (2.2)	91.2 (18.4)	12.9 (0.9)
	A340-541	52	11.7 (6.1)	11.0 (5.6)	13.1 (14.0)	4.7 (3.3)	78.5 (28.5)	14.6 (0.9)
	B767-300	91	5.8 (1.1)	8.9 (1.4)	21.0 (4.4)	2.5 (0.9)	98.4 (3.5)	17.4 (1.1)
	B777-300ER	131	23.0 (5.5)	21.0 (5.8)	23.1 (5.7)	6.7 (1.7)	90.5 (11.6)	32.2 (6.7)
Approach	A319-112	130	24.4 (8.2)	51.0 (9.9)	68.6 (21.9)	17.4 (6.8)	94.7 (4.9)	95.8 (13.5)
	A320-214	169	24.9 (5.9)	75.5 (16.3)	82.2 (20.5)	16.2 (5.1)	94.8 (4.0)	92.8 (14.0)
	A321-111	117	25.7 (7.4)	43.3 (14.8)	58.9 (26.0)	16.4 (4.4)	92.5 (5.4)	77.6 (16.0)
	A330-202	84	46.9 (13.8)	113.0 (22.6)	98.0 (25.8)	30.0 (11.2)	91.2 (7.5)	146.3 (26.5)
	A330-243	100	36.5 (6.5)	–	96.8 (25.8)	20.4 (10.3)	91.5 (8.1)	111.4 (23.1)
	A340-541	52	29.0 (10.2)	42.3 (14.0)	63.1 (20.1)	16.6 (2.6)	95.9 (3.5)	126.0 (24.2)
	B767-300	91	33.6 (6.3)	49.0 (13.2)	78.7 (29.3)	19.3 (7.1)	94.9 (6.1)	126.7 (37.7)
	B777-300ER	131	23.5 (7.3)	39.1 (9.2)	76.2 (23.6)	18.5 (4.7)	94.0 (4.7)	100.6 (16.8)

Note: Each entry shows the mean (and standard deviation within parentheses) of the evaluation metric across all the flights in the test data for that aircraft type. All the evaluation metrics are calculated on destandardized data (that is, data are retained at their original location and scale). The table also shows the total number of flights for each aircraft type in its FDR dataset. 'ICAO' refers to the ICAO-BFFM2 model. There are no entries for the A330-243 SFI models due to the model coefficients being unavailable.

and approach, respectively. The ICAO-BFFM2 method is used in AEDT to model the fuel flow rate in the LTO cycle when engine thrust values or other sources of the fuel flow rate are unavailable.

It is important to note that all the above APMs only provide point estimates of the fuel flow rate, and do not report any uncertainty estimates.

Evaluation of Fuel Flow Rate Models

Table 1 tabulates the predictive performance of the GPR models on the test datasets of the different aircraft types. Each entry shows the mean value (and the standard deviation within parentheses) of the evaluation metric across all the flights in the test dataset. Table 1 also reports the predictive performance of the BADA, SFI, ICAO-BFFM2 models on the test data.

It is observed that for all the aircraft types, the GPR models give a statistically significantly (at 5% significance level) lower mean absolute error than the other APMs. The median of the mean absolute errors across the different aircraft types in climb out is 3.8% for the GPR models, 10.7% for the BADA models, 10.2% for the SFI models, and 13.1% for the ICAO-BFFM2 models. In other words, the GPR models have an at least 62% lower median MAE than the other APMs in climb out. The median of the MAEs across the different aircraft types in approach is 18.0% for the GPR models,

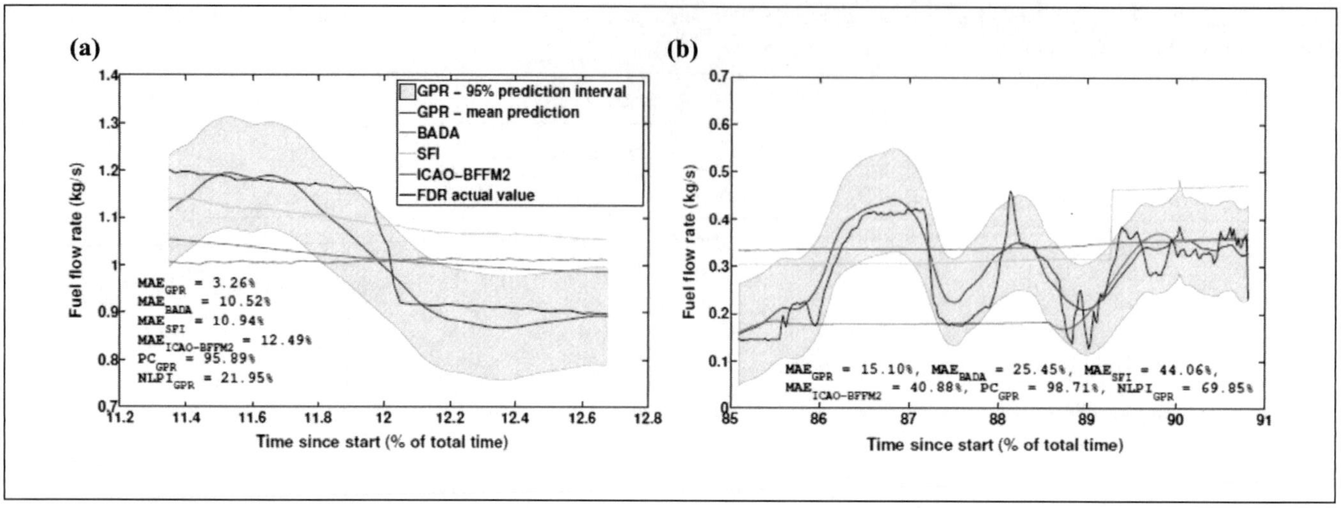

Figure 1. A321-111 fuel flow rate: Predictive performance of the Gaussian Process Regression (GPR), Base of Aircraft Data (BADA), Senzig-Fleming-Iovinelli (SFI), and International Civil Aviation Organization Databank-Boeing Fuel Flow Method 2 (ICAO-BFFM2) models for one test data flight in (*a*) climb out and (*b*) approach. The plots show destandardized fuel flow rates. The plots also show the actual fuel flow rate values from the Flight Data Recorder (FDR) dataset. The scales are different in the two figures. The values of the evaluation metrics (namely, Mean Absolute Error [MAE], Prediction Coverage [PC], and Normalized Length of Prediction Intervals [NLPI]) within the plots are the values for this particular flight alone.

27.4% for the BADA models, 49.0% for the SFI models, and 77.5% for the ICAO-BFFM2 models. The GPR models therefore, have an at least 34% lower median MAE than the other APMs in approach.

Unlike the other APMs which only report mean predictions of the fuel flow rate, the GPR models also give prediction intervals for the fuel flow rate. The median prediction coverage across the different aircraft types is 91.8% in climb out and 94.4% in approach, both of which are close to the nominal coverage probability of 95%. The median NLPI, indicative of the extent of relative variability or uncertainty in the fuel flow rate, across the different aircraft types is 17.9% in climb out and 106.0% in approach. The higher MAE as well as the larger uncertainty estimates in approach are due to the higher operational variability observed in this phase. In contrast to climb out, a typically smooth procedure, the fuel flow rate varies significantly in approach due to operational factors. These observations are illustrated in Figure 1, which shows the predicted and actual fuel flow rates in climb out and approach for one example flight of the A321-111.

Estimation of Fuel Mass Consumed

The fuel flow rate profiles estimated in climb out and approach can be further used to estimate the total mass of the fuel consumed (from all the engines) in climb out and approach. The equation for estimating the total fuel burn in a particular phase of flight is given by

$$\hat{m}_f = N_{\text{eng}} \sum_{i=1}^{n-1} \hat{\dot{m}}_{f\,i} \Delta t_{i+1}. \tag{17}$$

Here, $\hat{m}_f$ is the predicted total mass of fuel consumed in a particular phase of flight, n is the number of time instants in that phase of flight, $\hat{\dot{m}}_{fi}$ is the predicted fuel flow rate per engine for the ith instant, and $\Delta t_{i+1} = t_{i+1} - t_i$ is the time interval between the $(i+1)^{\text{th}}$ and the ith observations. At every instant, the fuel flow rate per engine predicted by a GPR model follows a Gaussian distribution. Therefore, the total mass of fuel consumed (divided by the number of engines and the total time in phase) follows a Gaussian Mixture distribution with mixture component weights governed by the time intervals between successive instants.

Table 2 shows the predictive performance of the total fuel mass estimation models on the entire unseen dataset (not used for training the fuel flow rate models) for the different aircraft types. In climb out, the GPR-based models give a median MAE of 2.0% across the different aircraft types. The median MAEs are 7.6%, 7.5%, and 10.1% for the BADA, SFI, and ICAO-BFFM2 models, respectively. A comparison of the mean absolute errors shows that, for all the aircraft types, the GPR-based models give a statistically significantly lower MAE (at a 5% significance level) than the other APMs. The median MAE in total fuel burn for the GPR-based models is more than 73% lower than the other APMs. The GPR-based models give a median NLPI of 30.6%, with a median PC of 100%. In the approach phase, the GPR-based models give a median MAE of 5.5% across the different aircraft types, compared with values of 13.8%, 42.8%, and 50.6% for the BADA, SFI, and ICAO-BFFM2 models, respectively. The GPR-based models give a

Table 2. Total Mass of Fuel Consumed: Predictive Performance of the GPR-Based Models and Other APMs in Climb Out and Approach for Different Aircraft Types

		Existing APMs			GPR models		
		Mean absolute error (%)				PC (%)	NLPI (%)
Phase	Aircraft type	BADA	SFI	ICAO	GPR	GPR models	
Climb out	A319-112	4.3 (3.5)	6.3 (3.6)	12.7 (4.9)	1.3 (1.3)	100.0 (0.0)	21.8 (3.6)
	A320-214	5.3 (4.4)	6.8 (2.8)	5.9 (4.2)	1.4 (1.1)	100.0 (0.0)	33.7 (5.6)
	A321-111	5.2 (3.2)	7.5 (5.6)	7.2 (5.5)	2.1 (1.5)	97.6 (15.6)	41.1 (8.1)
	A330-202	11.2 (2.0)	5.8 (2.1)	18.1 (3.1)	2.2 (1.7)	93.1 (25.8)	36.8 (4.4)
	A330-243	10.1 (2.1)	–	4.8 (3.3)	1.8 (1.2)	100.0 (0.0)	18.0 (2.9)
	A340-541	9.9 (5.7)	10.0 (5.3)	7.4 (11.8)	2.6 (3.1)	100.0 (0.0)	27.4 (7.1)
	B767-300	5.2 (1.4)	8.3 (1.8)	21.0 (3.6)	1.4 (1.3)	100.0 (0.0)	23.7 (4.6)
	B777-300ER	22.3 (6.5)	13.2 (7.0)	22.1 (7.1)	2.9 (2.1)	73.9 (44.4)	79.4 (9.0)
Approach	A319-112	13.7 (10.8)	42.8 (31.1)	53.3 (36.4)	3.9 (3.6)	100.0 (0.0)	133.6 (16.5)
	A320-214	13.8 (9.7)	63.5 (21.7)	55.9 (20.1)	5.5 (4.4)	100.0 (0.0)	133.6 (18.4)
	A321-111	15.5 (11.5)	31.3 (16.6)	45.8 (23.0)	4.9 (3.4)	100.0 (0.0)	114.3 (21.6)
	A330-202	11.8 (11.2)	82.9 (42.7)	47.8 (42.6)	6.2 (5.7)	100.0 (0.0)	188.9 (22.5)
	A330-243	14.7 (11.9)	–	85.8 (53.8)	7.4 (9.2)	100.0 (0.0)	150.7 (23.5)
	A340-541	24.5 (13.0)	55.2 (89.8)	72.6 (101.0)	5.4 (3.2)	100.0 (0.0)	160.4 (23.3)
	B767-300	13.3 (10.2)	25.7 (21.3)	31.3 (29.0)	4.6 (4.0)	100.0 (0.0)	167.6 (28.4)
	B777-300ER	13.2 (9.9)	22.1 (34.0)	38.1 (37.0)	5.9 (4.2)	100.0 (0.0)	153.2 (22.4)

Note: Each entry shows the mean (and standard deviation within parentheses) of the evaluation metric across all the flights in the test data for that aircraft type. All the evaluation metrics are calculated on destandardized data (that is, data are retained at their original location and scale). The table also shows the total number of flights for each aircraft type in its FDR dataset. 'ICAO' refers to the ICAO-BFFM2 model. There are no entries for the A330-243 SFI models due to the model coefficients being unavailable.

statistically significantly lower or similar MAE (at a 5% significance level) as the other APMs. The GPR-based models achieve a more than 59% reduction in the median MAE in approach. The GPR-based models again give a median PC of 100% in approach, though the extent of uncertainty is higher (as indicated by a median NLPI of 152.0%), which can be attributed to the higher operational variability in this phase.

Discussion and Next Steps

This paper has presented a methodology to statistically estimate aircraft fuel burn in the climb out and approach phases of flight, given trajectory data. Data from flight data recorders of a commercial airline were used for model building, making the models representative of real aircraft operations. The model features important for fuel flow rate estimation were determined using an understanding of the physical principles governing aircraft and engine performance. Modeling assumptions allowed the feature space to be restricted to trajectory variables easily obtainable/derivable from ground-based surveillance data sources.

A nonparametric probabilistic technique called Gaussian Process Regression (GPR) was used to model the fuel flow rate. Under the GPR formulation, the predictive distribution of the fuel flow rate at a new input is a Gaussian distribution, whose parameters (mean and standard deviation) are estimated. The predictive distribution can be used to determine the mean and the 95% prediction intervals for the fuel flow rate. The mean gives a point estimate of the fuel flow rate, while the prediction intervals quantify its uncertainty resulting from random disturbances, model assumptions, and unmodeled features, and the operational variability seen in fuel flow rates. The estimated fuel flow rates can be further used to determine the predictive distributions of the total fuel consumed in climb out or approach. The statistical models were compared with other commonly used APMs. It was found that the GPR-based models gave a better predictive performance for the fuel flow rates and the mass of fuel consumed.

The statistical models developed can be used to estimate the fuel burn and emissions impact of real flight trajectories, and default operational procedures. As a result, the user of these models can more realistically model fuel burn and emissions in the vicinity of airports. These models will help improve fuel burn and emission inventories by evaluating impacts on a per-flight operation basis (instead of on a fleet/airline/regional level alone). Our models estimate the fuel burn using trajectory information, thereby removing the need to estimate any hidden variables. This presents a significant advantage over other APMs, which often require an estimate of the net thrust, a challenging problem in itself. In contrast to other fuel burn models, statistical models provide a predictive distribution, rather than a point estimate. They therefore quantify the expected uncertainty in fuel burn; these uncertainty estimates can be used to identify anomalous behavior. More detailed discussions of these results can be found in (*24*).

The proposed models suffer from the common disadvantage presented by any data-driven model: They assume that the training dataset is representative of the range of operating points of interest. Therefore, they cannot be applied to input variable values that are radically different from those seen in the training data. It should be noted that the models developed in this paper were trained on a relatively small subset of flight data. Models trained on larger sets of data, covering a variety of operating environments, airports, airlines and aircraft types, would be more generalizable. Given such new training data, the methodology outlined in this paper can easily be used to develop new models. Data used for model building in this paper were not differentiated on the basis of airports. The impact of an airport on fuel burn can be assessed by developing airport-specific models. Similarly, additional features such as flight distances, and high-fidelity weather data may also be investigated. Finally, more precise estimates of the fuel mass consumed can be obtained by modeling it directly, rather than by aggregating fuel flow rate estimates.

Acknowledgments

This research was supported by the TRB through the 2016–2017 Graduate Research Award Program on Public-Sector Aviation Issues. The authors thank Monica Alcabin, Bob Arnott, Brian Kim, Larry Goldstein, and Saadat Syed for their insights and feedback, as well as Mary Sandy and Sarah Pauls of the Virginia Space Grant Consortium for logistical support. This research was also supported in part by the National Science Foundation under grant number 1239054, CPS:Frontiers:FORCES.

References

1. FAA Office of Environment and Energy. *Aviation Emissions, Impact, and Mitigation: A Primer*. FAA, Washington, D.C., 2015.
2. Owen, B., D. S. Lee, and L. Lim. Flying into the Future: Aviation Emissions Scenarios to 2050. *Environmental Science and Technology*, Vol. 44, No. 7, 2010, pp. 2255–2260.
3. GasTurb 12 software [Online], 2015. http://gasturb.de/. Accessed October 2015.
4. Senzig, D. A., G. G. Fleming, and R. J. Iovinelli, Modeling of Terminal-Area Airplane Fuel Consumption. *AIAA Journal of Aircraft*, Vol. 46, No. 4, 2009, pp. 1089–1093.
5. Trani, A. A., F. C. Wing-Ho, G. Schilling, H. Baik, and A. Seshadri. A Neural Network Model to Estimate Aircraft Fuel Consumption. *Proc., AIAA 4th Aviation Technology, Integration and Operations (ATIO) Forum*, Chicago, IL, 2004.
6. International Civil Aviation Organization. *ICAO Aircraft Engine Emissions Databank* [Online], 2014. https://easa.europa.eu/document-library/icao-aircraft-engine-emissions-databank. Accessed February 2014.
7. Chati, Y. S., and H. Balakrishnan. Analysis of Aircraft Fuel Burn and Emissions in the Landing and Take off Cycle using Operational Data. *Proc., 6th International Conference on Research in Air Transportation*, Istanbul, Turkey, 2014.
8. Enea, G., J. Bronsvoort, H. Fricke, C. Seiß, J. Rosenow, A. Ramadani, and M. Paglione. Fuel Burn Estimation Modeling for ATM Benchmark Applications Perspectives from an International Collaboration. In *USA/Europe Air Traffic Management Research and Development Seminar*, Seattle, WA, 2017.
9. Kulikov, G. G., and H. A. Thompson. (eds.). *Dynamic Modelling of Gas Turbines: Identification, Simulation, Condition Monitoring, and Optimal Control*. Springer-Verlag, London, 2004.
10. Chu, E., D. Gorinevsky, and S. P. Boyd. Detecting Aircraft Performance Anomalies from Cruise Flight Data. *Proc., AIAA Infotech @ Aerospace Conference*, Atlanta, GA, 2010.
11. Hastie, T., R. Tibshirani, and J. Friedman. *The Elements of Statistical Learning: Data Mining, Inference, and Prediction*. Springer-Verlag, New York, 2009.
12. Rasmussen, C. E., and C. K. I. Williams. *Gaussian Processes for Machine Learning*. The MIT Press, Cambridge, Mass., 2006.

13. Yoder, T. *Development of Aircraft Fuel Burn Modeling Techniques with Applications to Global Emissions Modeling and Assessment of the Benefits of Reduced Vertical Separation Minimums*. Master's thesis. Department of Aeronautics and Astronautics, Massachusetts Institute of Technology, Cambridge, 2007.
14. Cumpsty, N. A. *Jet Propulsion: A Simple Guide to the Aerodynamics and Thermodynamic Design and Performance of Jet Engines* (2nd ed.). Cambridge University Press, Cambridge, UK, 2003.
15. Dürichen, R., M. A. Pimentel, L. Clifton, A. Schweikard, and D. A. Clifton. Multi-Task Gaussian Process Models for Biomedical Applications. *Proc., IEEE-EMBS International Conference on Biomedical and Health Informatics*, Valencia, Spain, IEEE, New York, 2014.
16. Platt, J. C., C. J. C. Burges, S. Swenson, C. Weare, and A. Zheng. Learning a Gaussian Process Prior for Automatically Generating Music Playlists. In *Advances in Neural Information Processing Systems 14* (T. G. Dietterich, S. Becker, and Z. Ghahramani, eds.), MIT Press, Cambridge, MA, 2002, pp. 1425–1432.
17. Chati, Y. S., and H. Balakrishnan. A Gaussian Process Regression Approach to Model Aircraft Engine Fuel Flow Rate. *Proc., ACM/IEEE International Conference on Cyber-Physical Systems*, Pittsburgh, Pa., IEEE, New York, 2017.
18. Chati, Y. S., and H. Balakrishnan. Statistical Modeling of Aircraft Takeoff Weight. In *USA/Europe Air Traffic Management Research and Development Seminar*, Seattle, WA, 2017.
19. The Mathworks, Inc. *MATLAB® R2012a Documentation*. Software Package, 2012.
20. Vanhatalo, J., J. Riihimäki, J. Hartikainen, P. Jylänki, V. Tolvanen, and A. Vehtari. GPstuff: Bayesian Modeling with Gaussian Processes. *Journal of Machine Learning Research*, Vol. 14, 2013, pp. 1175–1179.
21. Nuic, A. *User Manual for the Base of Aircraft Data (BADA) Revision 3.13*. EEC Technical/Scientific Report No. 15/04/02-43. EUROCONTROL Experimental Centre, 2015.
22. Ahearn, M., E. Boeker, S. Gorshkov, A. Hansen, S. Hwang, J. Koopmann, A. Malwitz, G. Noel, C. Reherman, D. Senzig, G. B. Solman, Y. Tosa, A. Wilson, A. Zubrow, N. Didyk, J. DiPardo, F. Grandi, M. Majeed, J. Bernal, E. Dinges, D. Rickel, M. Yaworski, C. Hall, and S. Augustineetal />. *Aviation Environmental Design Tool (AEDT) Technical Manual, Version 2b, Service Pack 3*. Report No. DOT-VNTSC-FAA-16-11. U.S. Department of Transportation Volpe National Transportation Systems Center, 2016.
23. Herndon, S. C., T. Rogers, E. J. Dunlea, J. T. Jayne, R. Miake-Lye, and B. Knighton. Hydrocarbon Emissions from In-Use Commercial Aircraft during Airport Operations. *Environmental Science and Technology*, Vol. 40, No. 14, 2006, pp. 4406–4413.
24. Chati, Y. S. *Statistical Modeling of Aircraft Engine Fuel Burn*. Ph.D. thesis. Department of Aeronautics and Astronautics, Massachusetts Institute of Technology, Cambridge, MA, 2018.

The ACRP Selection Panel for the Graduate Research Award Program on Public-Sector Aviation Issues peer-reviewed this paper (18-01539).

TRR
JOURNAL OF THE TRANSPORTATION RESEARCH BOARD

Article

Transportation Research Record
2018, Vol. 2672(29) 12–22

DOI: 10.1177/0361198118799711
journals.sagepub.com/home/trr
SAGE

Airport Expansion and Endangered Bats: Development and Mitigation Actions Near the Indianapolis International Airport

Timothy J. Divoll[1] and Joy M. O'Keefe[1]

Abstract

Economic prosperity and globalization are major drivers for development of international airports, but aviation-oriented businesses and residential developments are a by-product of airport business models. Among the multitude of planning and development considerations is the habitat needs of endangered wildlife species. Foraging data were analyzed from 57 bats during three time periods (1998–1999: pre-mitigation; 2005–2006: during mitigation, and 2014–2016: post-mitigation) of a long-term study of Indiana bats (*Myotis sodalis*) near the Indianapolis International Airport. At this site, both developed land cover and forested land cover increased between 1998 and 2016 (34.1% and 3.3%, respectively). Mitigation actions included converting 323 ha of residential lots back to forest, and creation of a 56 ha wetland and an 85 ha multi-use park. Bat use of landscape cover types was related to changes in land cover during each period and competing hypotheses were compared to explain changes in bat foraging space use. With the addition of a major highway interchange where the colony foraged, bats increased foraging ranges, presumable in search of new habitat. In all periods, bats selected for forested habitat; as trees in replanted forest and designated parks aged, bats reduced their foraging ranges. Restoring hardwood forest and setting aside parklands were effective proactive mitigation measures for the colony of Indiana bats near the Indianapolis International Airport, and similar actions should benefit other wildlife where human development and habitat needs intersect.

Landscape change is a necessary consequence of human progress and development. Some changes happen rapidly and drastically, such as large-scale land clearing, and others are less obvious, such as slow commercial or residential fill-in over several decades. Economic prosperity and globalization are major drivers for development of international airports (*1*), and aviation-oriented businesses and residential developments are a by-product of airport business models (*2*). Increased traffic into and out of airports requires more pavement, potentially at the expense of wildlife habitat as the developed footprint expands. Runways must be maintained and surrounding vegetation must be kept clear for air traffic safety, while managing for potentially hazardous wildlife that may impact air traffic (*3*). Among the multitude of planning and development considerations are the habitat needs of endangered wildlife species. Biological assessments and consultations with the United States Fish and Wildlife Service (USFWS) are required under the Endangered Species Act (ESA) for airport and other development projects to proceed when endangered species may be affected. For some species, such as Indiana bats (*Myotis sodalis*), detection may be biased toward point locations such as roosting sites, whereas functional habitat needs may be far greater than apparent (i.e., foraging habitat). This phenomenon has been demonstrated with highly mobile endangered swift parrot migrations in Tasmania; space use estimates varied drastically each year depending on resource availability (*4*).

Estimating space use and habitat selection are important aspects of airport development planning and satisfying ESA requirements. For example, identifying endangered burrowing owl burrow locations at the Mineta–San Jose Airport in California required active management on airport properties to ensure no owls were injured or killed (*5*). In that study, managers were able to provide artificial burrows away from runways to

[1]Center for Bat Research, Outreach, and Conservation, Department of Biology, Indiana State University, Terre Haute, IN

Corresponding Author:
Address correspondence to Timothy J. Divoll: tdivoll@sycamores.indstate.edu

safely maintain habitat needs of the owls. In many situations, wildlife researchers and managers typically implement radio telemetry as a means to gather information about the movement patterns and foraging habits of wild animals. Telemetry data may then be used to estimate space use and habitat selection as a function of resource utilization (*6*).

Bats are particularly challenging animals to study, flying at speeds up to 17 ms^{-1} and largely unrestricted by landforms (*7*). Because of their small size, they pose little collision risk to aircraft (*8*). However, their small size and mobility spur unique data collection challenges—bats continuously move while workers try to triangulate locations, and the window of opportunity to collect movement data with miniature radio tags is limited. Researchers typically end up with <1 week of data on each individual, representing a snapshot of foraging activity during a narrow temporal window rather than a robust picture of the entire habitat footprint. Snapshots of data make it difficult to assess preference or selection for certain habitat types and the entirety of space used. These difficulties hamper development planning and conservation decisions.

Many animal movement studies account for uncertainty using kernel density estimates to create a continuous probability of use with a home range, and then relate the area used to underlying habitats with a resource utilization function model (*9*), distance-based analyses (*10, 11*), or compositional habitat analysis (*12*). All of these methods have shed light on bat habitat preferences and space use, but results vary from study to study, making it difficult to glean species-specific management and conservation implications. Results of these methods may be sensitive to strong individual bat habitat preferences, making it difficult to generalize and apply to other species or landscapes. Womack et al. found great heterogeneity in individual resource selection within a colony of Indiana bats in Missouri (*13*), which may be important when considering the needs of an entire colony. Most habitat studies of bats last less than three years and none have considered bat response to direct mitigation actions (*14*). Thus, most conservation decisions for bats are informed by cumulative evidence from many localized studies for short periods of time, making it impossible to discern how bats tolerate long-term changes on the landscape.

During the growth of Indianapolis International Airport (IND) over the past 20 years, its land acquisition needs overlapped with a large colony of endangered Indiana bats (*15*). For development to proceed, conservation actions were necessary to ensure persistence of the bat colony. From mitigation actions sparked by agreements between the Indianapolis Airport Authority and USFWS, the opportunity now exists to evaluate the effectiveness of the actions imposed. There was no prior evaluation of the colony's ability to cope with changes on the landscape, and the actions were assumed to be a success at the time of implementation. Such a long-term dataset holds value for airport expansion and land management as well as an understanding of how tolerant endangered bats are to changing landscapes. A large part of the mitigation requirement was an annual survey of the colony's roosting and foraging activities during the summer maternity season. The colony of Indiana bats at IND has persisted despite major landscape changes—some habitat was lost, but the bats continue to use dedicated wetlands, parks, and forests set aside as mitigation to offset spatial expansion and development at IND. In addition, over 3,000 artificial roosting sites were erected to provide roosting options as roost trees became more limiting with forest loss (*16*). Because bats travel from roosting sites to foraging areas (*17*), a landscape approach was taken in this analysis with a focus on foraging area requirements in relation to landscape changes.

The objective was to analyze changes in foraging space use over time as development encroached on existing habitat, and as mitigation plantings aged and became more suitable habitat. As part of the mitigation plan of this project, 18 years of foraging data were collected (1997–2016) from 141 individual bats. A subset of data was analyzed from 57 bats tracked in three critical time periods: pre-mitigation (1998 and 1999); during mitigation (2005 and 2006); and post-mitigation (2014–2016). It was hypothesized that bats would increase their foraging ranges as development replaced forested habitat. Specifically, it was predicted that space use would increase over time with increased development under the assumption that bats would need to venture farther from roosts to find quality habitat. This case study was used to develop a generalized linear mixed-effects model broadly applicable to any situation where development and wildlife habitat needs intersect. Because wildlife habitat use data rarely capture all of an animal's space use, sampling error and reproductive status were incorporated to account for year-to-year sampling biases and differential energetic demands.

Methods

Between the early 1990s and mid-2000s, the Indianapolis International Airport transitioned from a small regional airport to a major freight hub with international connections, which required construction of new terminals and runways. On detection of a colony of endangered Indiana bats in 1992, USFWS issued a Biological Opinion in regard to an incidental take proposal for runway expansions that would remove some bat habitat. Indiana bats were one of the first species listed in 1967 because of disturbance at their limited winter hibernacula (*18*). Initial mitigation (1992–1999) included land parcel acquisition, wherein a large 56 ha wetland was created in the middle of the study area and 323 ha of residential lots were converted back to forest with hardwood plantings (seedlings) (*19*). Thereafter, annual monitoring of the bat colony ensued with monitoring plans through 2016 (*19*). In 2002, development of the Six Points Interchange on I-70 required additional years of monitoring under a Habitat Conservation Plan. The interchange facilitated rapid access to the airport and made commercial and residential development more attractive in the immediate area, but this major landscape change required rerouting the East Fork of White Lick Creek (*20*) and thoughtful clearing of potential roost trees (*21*). In 2008, FedEx established a regional hub, which sparked further development of associated fulfillment centers from large companies such as Amazon. These expansion activities brought significant economic benefits to the state and were only possible with unique and proactive measures implemented to offset habitat disturbance. Although initial mitigation began in 1992, the focus herein was on bat response to the second round of mitigation related to the Six Points Interchange.

Collecting Data on Space Use

As part of mitigation monitoring, 6–12 Indiana bats were tracked by graduate students and technicians each year during 1998–1999 and 2002–2016 for a total of 141 individual bats. The major goal was to estimate the overall foraging footprint and spatial juxtaposition to development and mitigation areas. Miniature radio transmitters weighing less than 0.5 g ($\leq$5% of body weight) were glued to the backs of bats, and teams of 3–6 observers triangulated foraging locations over 1–7 nights. A minimum of 30 relocations were necessary to include an individual in the analysis, and up to 380 data points were collected from any one bat. Although the team of trackers varied each year, information about study site geography and the tracking protocol were bestowed to each successive graduate student. Relocations were estimated for a subset of 57 adult females (n = 9 pregnant; n = 19 lactating; n = 20 post-lactating) or juveniles (n = 9) from three critical time periods: pre-mitigation (1998 and 1999); during mitigation (2005 and 2006); and post-mitigation (2014–2016). Maximum likelihood estimators were used to pinpoint relocations in LOAS 4.0 software (*22*), kernel density estimates were created with a cross-validated (CVh) smoothing parameter in the Geospatial Modeling Environment (*23*) and continuous foraging ranges generated for each bat with $\geq$30 relocations, interpolating spatial areas between relocated points. The total area within each bat's foraging range was then measured in hectares (ha). Annual observer error was estimated using the ratio of attempted to successful triangulations to account for year-to-year differences in tracking team abilities. It was also expected that bats with more exploratory foraging habits were more difficult to obtain data on, thereby increasing the amount of error in relocation estimation.

Analyzing Habitat Selection

Next individual bat foraging density estimates were brought into ArcGIS v10.1 to extract underlying habitat use data. To do this, annual habitat maps were created by classifying National Agriculture Statistics Survey layers into the following four categories: water/wetland, developed (commercial and residential), agricultural, and forest. The study area was first defined by plotting all bat foraging relocations from 1998, 1999, 2005, 2006, 2014, 2015, and 2016; annual habitat layers were then clipped by the extent of the outermost points (~11.5 km east–west by 13 km north–south; 15,000 ha; Figure 1). The values of each habitat type were summed to estimate the amount of habitat available to the bats each year. Foraging density values for each bat were then summed within each of the habitat categories to estimate the proportions of each habitat used by each bat; higher density foraging areas were more important (*24*). Within each time period of the study (pre-mitigation, during mitigation, post-mitigation), overall habitat selection was tested for with a multivariate analysis of variance (MANOVA; 1,000 randomizations, α = 0.05) and then the importance of each habitat was ranked using log-odds ratios (*25*). This foraging data set is one of the largest and longest term of its kind, which retrospectively was only possible because of the Indianapolis Airport Expansion projects.

Modeling Bat Response to Landscape Changes

The 'lme4' package (*26*) in program R (*27*) was used to build linear mixed effect models, and seven competing models were compared with an information theoretic approach (*28*) to determine how bats altered their space use in response to changes on the landscape. It was predicted that changes in land cover types over time were related to changes in foraging behavior and, thus, the amount of space (in ha) that bats required to attain foraging resources. The response variable, space use, was log-transformed to normalize data that were skewed toward smaller home ranges (mean = 600 ha) despite several bats using >1200 ha. All models included percent forested habitat used and percent developed used as fixed effects, with year and relocation success rate as random effects; the null model only included random effects (Table 1). More complex models were developed with

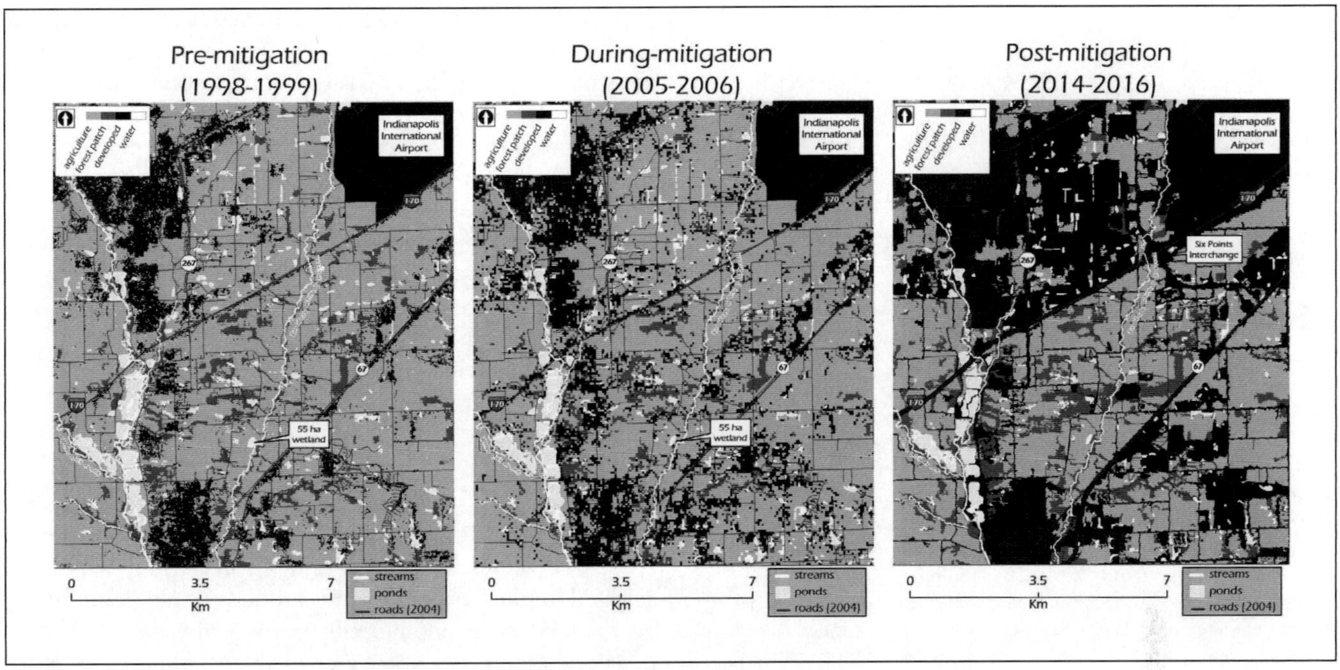

Figure 1. Habitat change near Indianapolis International Airport, 1998–2016.
Comparison of habitat available at the beginning (1998), middle (2006) and end (2016) of a long-term study. Both developed and forested land cover types increased whereas agriculture decreased over time. Classification of mitigation study periods was specific to the Six Points Interchange and rapid commercial development after 2003. The Dynamap/2000® roads layer was used from 2004 for each study period.

Table 1. Mixed Model Rankings

Model	w_i	Δ_iAIC
forest_used + devel_used + year + err_rate	**0.63**	**0.00**
forest_used + devel_used + year + err_rate + forest_avail	**0.27**	**1.68**
forest_used + devel_used + year + err_rate + repro	0.07	4.52
forest_used + devel_used + year + err_rate + forest_avail + repro	0.03	6.13
forest_used + devel_used + year + err_rate + forest_avail + repro + forest_used*repro	0.01	9.72
forest_used + devel_used + year + err_rate + repro + forest_used*repro	0.00	11.42
null (year + err_rate)	0.00	18.88

Note: Model weights (w_i) and differences in Akaike Criterion Score (ΔAIC) for competing models explaining factors related to behavioral changes in Indiana bat (*Myotis sodalis*) space use during three phases of mitigation between 1998 and 2016: pre = pre-mitigation (1998–1999), during = during mitigation (2005–2006), post = post-mitigation (2014–2016). Classification of mitigation study periods was specific to the Six Points Interchange and rapid commercial development after 2003. Plausible models (≤2 ΔAIC) are shaded bold. forest_used = average percentage of forest used; devel_used = average percentage of developed habitat used; year = year of sampling; err_rate = successful/attempted relocations; forest_avail = percentage of forest available in study area; repro = reproductive condition of bat.

the amount of forested habitat available, bat reproductive condition, and the interaction between forested habitat used and reproductive condition to account for changes in space use related to energetic demands or loss of habitat. Because conditions such as pregnancy and lactation (*29*) or naivety (e.g., juveniles [*30*]) may affect foraging behavior, reproductive status of adult females and non-reproductive juveniles were considered as factors. Available forested and developed areas were positively correlated ($r = 0.86$; $p < 0.001$), so only forest availability was included in models as it is the preferred habitat for bats. The amount of water available and used changed very little during the study period, so water was not included in the models. Open habitat used was negatively correlated with forested habitat used ($r = -0.71$; $p < 0.001$), therefore open habitat was not included in the models. Models within 0–2 Δ Akaike's Information Criterion (AIC) scores were considered equally plausible to explain the observations (*28*). AIC scores were weighted and models considered in a confidence set if

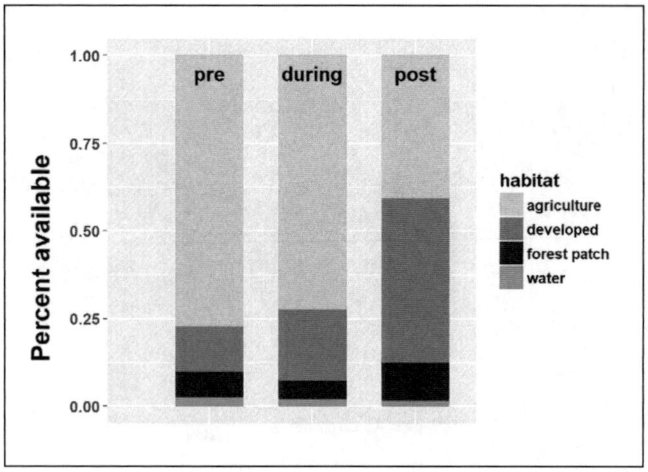

Figure 2. Habitat available during each time period.
Percentages of habitat available at critical periods of a long-term study: pre = pre-mitigation (1998–1999), during = during mitigation (2005–2006), post = post-mitigation (2014–2016). Classification of mitigation study periods was specific to the Six Points Interchange and rapid commercial development after 2003. Over the course of the study, agricultural areas were converted to developed land or converted back to forest patches.

weighted scores were within 10% of the top model. Fixed parameters were then weighted within the confidence set by unconditional variance to calculate 85% confidence intervals (CI [*28*]). When upper and lower CI values did not cross zero, those parameters carried enough weighted evidence to be important factors related to bat space use.

Results

During the study period, agricultural cover declined by 36.5% but developed land cover increased by 34.1%. In addition, water decreased by 0.8% and forest patches increased by 3.3% (Figure 1). The most drastic changes in composition of habitat were at the end of the study period; both developed area and forest patches increased over time as a result of airport progress and associated mitigation actions (Figures 1, 2).

Space Use Over an 18-year Period

During the pre-mitigation period, bats (n = 13) foraged predominantly in forest patches north and south of I-70, including the area that would soon be developed for the Six Points Interchange, using a total area of 3,739 ha (Figure 3). After development of the interchange in 2003, bats (n = 20) displayed what may have been exploratory behavior to the north along the East Fork of White Lick Creek and to the west near a series of quarry ponds, using a total of 6,591 ha (Figure 3). In the post-mitigation phase, bats (n = 24) reduced foraging north of I-70, but continued to forage in forested patches at the heart of the study area and the quarry ponds to the west, using a total area of 4,019 ha (Figure 3). Individual adult female bats used less than 990 ha each in the pre-mitigation period, but showed more exploratory foraging behavior by nearly doubling their space use in the middle of the study (up to 3,063 ha for one bat; Figure 4), regardless of reproductive condition (Figure 5). As forest plantings aged in the post-mitigation period, individual adult female foraging ranges decreased to a maximum of 859 ha (Figure 5). Individual juvenile bats used less than 920 ha at all phases of mitigation (Figure 5).

Habitat Selection Changes

In all three time periods, bats selected for forested habitat over agriculture, water, and developed land ($p \leq 0.002$; Table 2). In the pre-mitigation period, bats also selected agriculture over water and developed land; however, in later periods agriculture was only selected over developed land and not water (Table 2). Thus, bats preferred forested habitat and secondarily used agricultural areas despite changes in habitat availability (Figure 2) and space use (Figure 5) in each study period.

Factors Related to Behavioral Shifts

The top model included forest used, developed land used, year, and error rate. The next highest ranked model, which also included forest availability was also plausible (Table 1). Percent of forest used negatively influenced space use; as the amount of forest use increased, space use decreased. On average, the foraging densities of bats in the pre-mitigation period were comprised of 30% forest (n = 13 bats), whereas foraging densities contained 19% forest during mitigation (n = 20 bats) and 31% post-mitigation (n = 24 bats). The percent of developed land used and the amount of forest available did not carry enough weighted evidence to influence space use by bats, with 85% CIs crossing zero (Table 3). As bats adhered to forested areas, they required less space (Figure 4); as they ventured across a greater proportion of developed and open agricultural land, they traveled farther in search of foraging resources (Figure 3), regardless of reproductive condition (Table 1; Figure 5).

Discussion

This case study provides an example of evaluating mitigation actions intended to offset the consequences of habitat removal. Although initial efforts focused on roosting habitat and monitoring (*15*), the greater space requirements for foraging proved to be an important consideration in this dynamic between an endangered

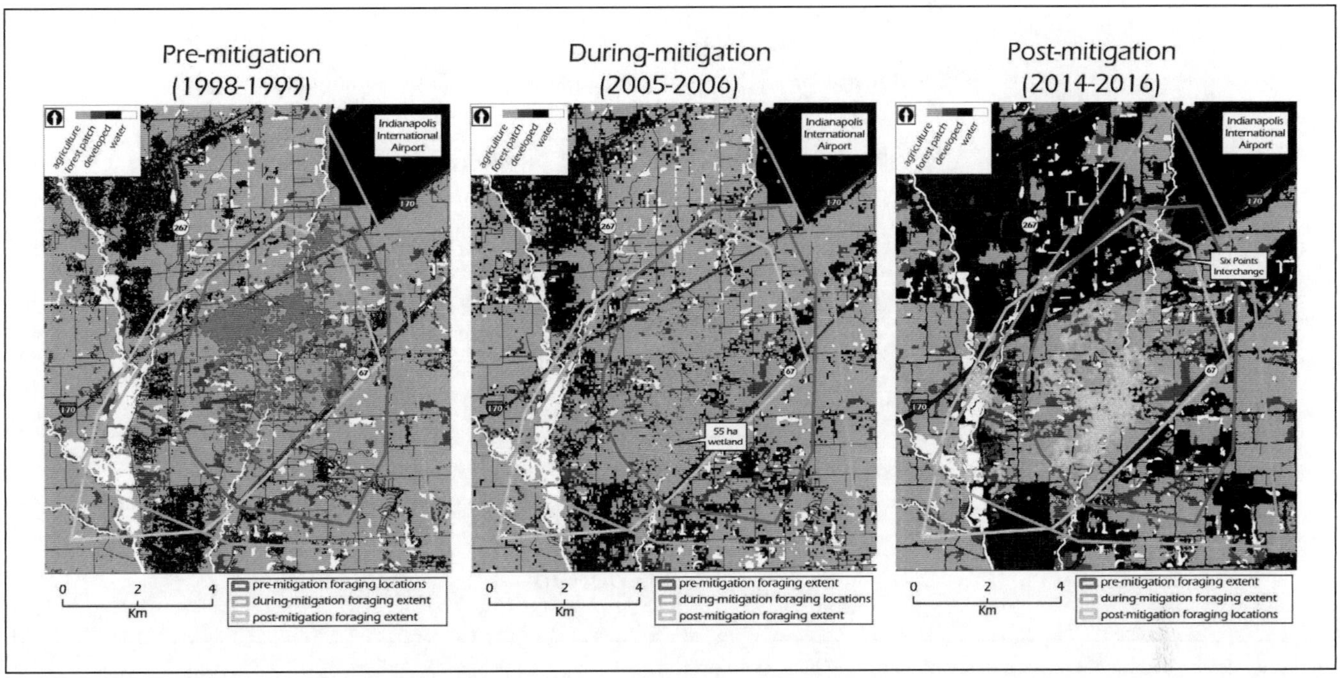

Figure 3. Bat foraging activity in proximity to the Indianapolis International Airport.
Comparison of foraging activity at the beginning (1998), middle (2006) and end (2016) of a long-term study. The Indianapolis International Airport area is markedly more developed at the end of the study, reflecting a transition from a regional airport to an international and freight hub. Foraging locations within critical period (pre-, during-, or post-mitigation) are displayed as points whereas foraging extents for alternate critical periods are displayed as minimum convex polygons. Classification of mitigation study periods was specific to the Six Points Interchange and rapid commercial development after 2003.

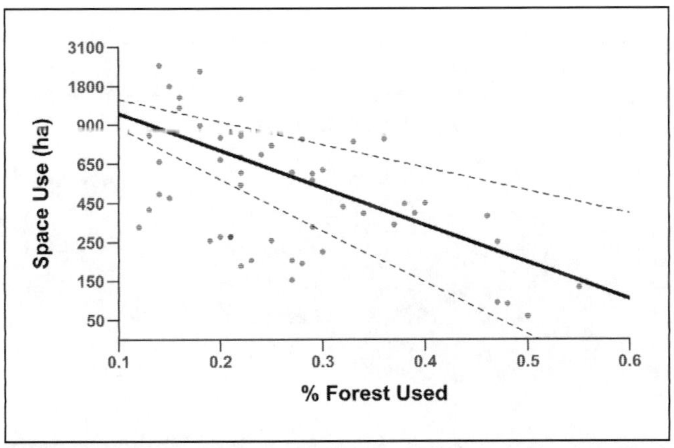

Figure 4. Model-averaged relationship between space used and forest used.
Space use in hectares (ha) dependent on the proportion of forested habitat used by 57 adult female (n = 9 pregnant; n = 19 lactating; n = 20 post-lactating) and juvenile (n = 9) Indiana bats (*Myotis sodalis*) between 1998 and 2016 at the Indianapolis International Airport. Space use data represent log-transformed values, percent forest used represents the weighted proportion of forested habitat within a bat's foraging range. Solid line = model-averaged regression; dashed lines = bounds of 85% upper and lower confidence intervals.

species and human development. As the amount of agricultural land decreased near IND, developed land replaced it, following aerotropolis business model predictions (*3*); however, in the heart of the bat colony's foraging area, a small percentage of agriculture was converted to forest (Figure 3), with replanted areas connecting larger existing forest patches (Figures 1, 3).

As habitat availability in the study area changed between time periods (Figures 1, 2), bats focused foraging activities near forested areas, using less space as they used a greater proportion of forest (Figure 5). Initially, quality foraging habitat was found along the East Fork of White Lick Creek and nearby forest (Figure 3). As development ensued, bats ventured out in search of new foraging habitat (Figures 3, 5) until mitigation plantings and dedicated parks became presumably more suitable foraging habitat near the end of the study (Figure 3). Overall foraging activity in the post-mitigation phase was largely restricted to the area between I-70 (interstate highway) and Route 67 (4-lane major artery); the bat colony persisted and bats continued to select forested habitat (Table 2) despite the large increase in developed land (36.5%) and the relatively small increase in forested land (3.3%) (Figures 1, 2). These observations largely agree with observations of Indiana bat foraging behaviors in Indiana and other locations. Sparks et al. initially analyzed habitat use from 11 female bats tracked in 2005 at IND and found that bats selected for forested woodlands, followed by agriculture and open water (*12*). Within a similar landscape in Illinois, Menzel et al. found

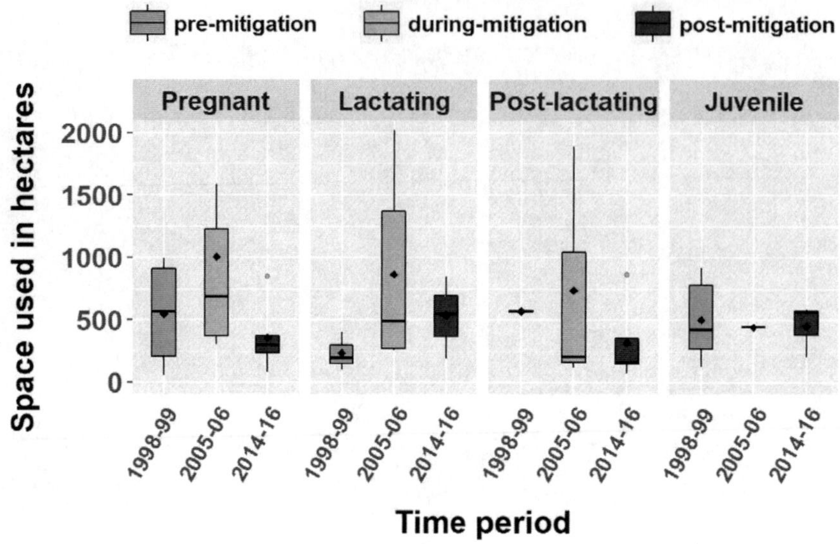

Figure 5. Space use by bat reproductive condition in each time period.
Space use (ha) by 57 adult female (*n* = 9 pregnant; *n* = 19 lactating; *n* = 20 post-lactating) and juvenile (*n* = 9) Indiana bats (*Myotis sodalis*) at varying stages of development and mitigation at the Indianapolis International Airport. Classification of mitigation study periods was specific to the Six Points Interchange and rapid commercial development after 2003. Box plots represent the median (solid bar), mean (diamond), and 25th and 75th quartiles with standard error bars. One pregnant bat in 2005 had a foraging range of over 3,000 ha; she is included in the descriptive group parameters but was left off of the figure to better display the axes.

Table 2. Habitat Selection Rankings per Time Period

Study period			Years; sample size	
Pre-mitigation			(1998–1999; *n* = 13 bats)	
Wilk's lambda = 0.0437; *p* = 0.002				
	forest	agriculture	water	developed
forest	0	+ + +	+ + +	+ + +
agriculture	− − −	0	+ + +	+ + +
water	− − −	− − −	0	+ + +
developed	− − −	− − −	− − −	0
During mitigation			(2005–2006; *n* = 20 bats)	
Wilk's lambda = 0.0753; *p* = 0.001				
	forest	agriculture	water	developed
forest	0	+ + +	+ + +	+ + +
agriculture	− − −	0	+	+ + +
water	− − −	−	0	+
developed	− − −	− − −	−	0
Post-mitigation			(2014–2016; *n* = 24 bats)	
Wilk's lambda = 0.0557; *p* = 0.001				
	forest	agriculture	water	developed
forest	0	+ + +	+ + +	+ + +
agriculture	− − −	0	+	+ + +
water	− − −	− − −	0	+
developed	− − −	−	−	0

Note: Habitat selection for each time period of the study, ranked from log-odds ratios on all pairwise comparison within time period with a MANOVA (1000 randomizations, α = 0.05). Overall selection was determined by the Wilk's lambda statistic and significant differences among habitats are shown with either + + + or − − − signs; otherwise signs indicate non-significant relationships. In all periods, Indiana bats (*Myotis sodalis*) selected forest over other land cover types. Classification of mitigation study periods was specific to the Six Points Interchange and rapid commercial development after 2003

Table 3. Coefficient Estimates for Fixed Parameters in Top Models

Parameter	Estimate	SE	Upper 85% CI	Lower 85% CI
Intercept	3.25	0.27	3.70	2.80
Percent of forest used	−1.90	0.44	−1.17	−2.64
Percent of forest available	−0.33	1.13	1.56	−2.21
Percent of developed used	−0.01	1.00	1.66	−1.68

Note: Model-averaged parameter estimates, standard error (SE) and 85% confidence intervals in top-weighted models explaining space use by Indiana bats (*Myotis sodalis*) at the Indianapolis International Airport at critical periods of a long-term study: pre = pre-mitigation (1998–1999), during = during mitigation (2005–2006), post = post-mitigation (2014–2016). Classification of mitigation study periods was specific to the Six Points Interchange and rapid commercial development after 2003

Indiana bats selecting for roads, forest, riparian, and agricultural areas, in that order (*31*); in the current study roads were combined with developed areas because a previous study found an aversion to roads near IND (*32*). Bats selected forested habitat over developed land through the duration of this study (Table 2), despite losing forested habitat north of I-70. The colony ultimately abandoned areas to the north and shifted foraging activity to the south and west of IND (Figure 3).

One factor that was difficult to account for in this analysis was the 55 ha wetland (classified as forested habitat) created as mitigation habitat in the center of the study area, which was already present in 1998, the year that foraging data collection commenced (Figure 1; *19*). Small Myotid bats, such as Indiana bats, prey on insects associated with water and wetlands (*33, 34*), therefore the newly constructed wetland near IND likely attracted foraging bats before 1998, the first year of this study. Bats may have habituated to this mitigation action before the sampling window, masking the study's ability to detect a response. Indiana bats at IND did select for agricultural (i.e., open) habitat more than water and developed land in the pre-mitigation phase, but the new wetland was concealed by tree cover (*19*) and, thus, this potentially critical water resource was likely undervalued in the analysis. However, the pattern of foraging range exploration in the during-mitigation phase shows that bats were traveling some distance away from the large wetland in search of additional habitat (Figure 3). As of 2015, the colony of bats at IND mostly used artificial roosting boxes in the center of the study area (*35*), which likely allowed the colony to return each summer despite changes on the landscape. The importance of roosting structures was not investigated, but all roosting areas were contained within colony foraging areas. Further aspects of bat ecology, such as the interaction between roost sites and foraging areas, may influence planning and decision-making.

There are many potential factors other than available habitat involved in foraging decisions, thus affecting space use in each time period of the study. Weather, predation risk, or artificial light may also influence bat foraging decisions (*36–38*). It is expected that artificial light at IND increased with increasing development, which may have been a factor in the abandonment of habitat to the north of I-70 near IND. In Australia, Scanlon and Petit found most bat species avoiding lit areas but traveling into dark urban parks to forage (*37*). In another study, in Italy, Russo et al. found that increased light pollution had a stronger negative effect on bat drinking activity than on foraging (*39*). In addition to ensuring that mitigation actions cover the full extent of bat habitat (e.g., roosting, foraging, commuting, drinking), dark drinking sites away from light pollution should also benefit all wildlife species. As part of the 2002 HCP for development of the Six Points Interchange, an 85 ha multi-use park was created (Sodalis Nature Park; forested trails with a 3.5 ha pond) in the center of the mitigation area, and this park remained an important area for roosting and foraging into 2016 (*35*). The creation of forested patches and designated multi-use parklands in the urban-agricultural matrix near IND (i.e., forested bat habitat) were the most important mitigation actions in this study. For the most part, core use areas of Indiana bats in this study adhered to mitigation areas and remaining forest (Figure 3). Even a 3.3% increase in forested habitat was enough for the colony to remain at IND despite drastic landscape changes related to airport, commercial, and residential development.

Because bats are highly mobile when foraging at night, their total space use was likely underestimated. Using density of foraging and a weighted compositional approach to assess habitat selection allows for the inclusion of individual heterogeneity in behaviors but gives more importance to areas visited more often when calculating habitat use. A similar approach can be adopted for other wildlife inhabitants near major commercial or industrial sites to more accurately determine habitat selection. Regardless of wildlife taxa, defining the extent of space needed and core use areas is critical to maintenance of habitat contiguity when designing development projects or considering restoration areas, as recently demonstrated with Florida Scrub-Jays at the Kennedy Space Center, where intensive management in core, support, and auxiliary habitats aims to double the population of this listed species (*40*). At IND, Indiana bats required a total foraging area of 4,019 ha, but a core use area of 2,300 ha (Figure 3). Restoration or mitigation efforts should first focus on core use areas and then supporting habitat to connect important areas—recognizing that the shape and size of these requirements will be different depending on the wildlife species. Monitoring programs, like the one required for Indiana bats at IND, are necessary early on in development plans to determine space use requirements for sensitive wildlife.

As future transportation business models transition from large hub and spoke transportation centers to many smaller sites that support vertical take-off and landing (VTOL) for personal transport (*41*), the effects of increased air traffic in relation to sensitive wildlife species will need to be assessed. With more regional, but smaller developments, there is greater potential for compounding effects, such as farther-reaching artificial light (*38*), which may fragment natural habitats and affect species in different ways. Although VTOL sites may have small spatial footprints, landscape-scale efforts focused on alternative planning and mitigation actions may be necessary if residential and commercial development increases near numerous rural VTOL sites. Noise did not appear to cause Indiana bats at IND to abandon the study area, but other wildlife species may be more sensitive to airport-associated noise.

Little brown bats (*Myotis lucifugus*) and tri-colored bats (*Perimyotis subflavus*) are species similar to, but far more widespread than, Indiana bats, and populations of both have recently become imperiled because of disease (*42*). Remaining forested habitat near water is critical for these species (*43, 44*), but information on space use by individuals or a colony near or in a developed landscape is lacking. Because these bat species are widespread across most of North America (*45, 46*) and commonly found in fragmented or edge-dominated habitats (*47*), there is great potential for competing interests with development. A recent Prelisting Conservation Policy (PCP; Director's Order No. 218; https://www.fws.gov/endangered/improving_ESA/prelisting-conservation.html) was implemented as a proactive measure to mitigate for at-risk species not yet listed under the ESA. Under the PCP, credits gained from proactive mitigation may be later redeemed or sold if the target species becomes listed. Such approaches may benefit species such as the little brown bat in a manner similar to the unique agreements and proactive mitigation efforts between the Indianapolis Airport Authority and the USFWS. In a sense, bats with their large foraging ranges, may act as umbrella species to protect the habitat needs of other species during forest restoration efforts. In Sweden, restoration efforts for the white-backed woodpecker resulted in habitat creation for imperiled beetle species (*48*); likewise, restoring bat habitat near developed areas may also protect other sensitive species.

In this study, a unique data set was gathered that allowed quantification of foraging space use in relation to landscape changes. Restoring hardwood forest patches and setting apart parklands were effective proactive mitigation measures for the colony of Indiana bats near the Indianapolis International Airport. In addition to protecting a colony of endangered bats, mitigation actions also provided green space for recreation in a rapidly developing area as well as additional forested habitat and water resources for various other wildlife species near a major urban center.

Acknowledgments

Funding for this research was provided by the Airport Cooperative Research Program at The National Academy of Sciences and the Virginia Space Grant Consortium, in conjunction with the Transportation Research Board. We thank the following panel members for development and editing of this manuscript: Larry Goldstein, Sarah Pauls, Mary Sandy, Michael Lamprecht, Linda Howard, Natalia Alexandrov and Monica Alcabin. We thank countless graduate students and technicians that collected bat relocation data from 1998–2016. In particular, Dale W. Sparks and John O. Whitaker Jr. led monitoring efforts up until 2012. Lori Pruitt provided thoughtful comments on an earlier version of this manuscript.

References

1. Freestone, R. Planning, Sustainability and Airport-Led Urban Development. *International Planning Studies*, Vol. 14, No. 2, 2009, pp. 161–176. http://dx.doi.org/10.1080/13563470903021217.
2. Kasarda, J. D., and S. J. Appold. Planning a Competitive Aerotropolis. In *The Economics of International Airline Transport*, Emerald Group Publishing Limited, Bingley, United Kingdom, 2014, pp. 281–308.
3. Blackwell, B. F., T. L. DeVault, E. Fernández-Juricic, and R. A. Dolbeer. Wildlife Collisions with Aircraft: A Missing Component of Land-Use Planning for Airports. *Landscape and Urban Planning*, Vol. 93, No. 1, 2009, pp. 1–9. http://dx.doi.org/10.1016/j.landurbplan.2009.07.005.
4. Webb, M. H., A. Terauds, A. Tulloch, P. Bell, D. Stojanovic, and R. Heinsohn. The Importance of Incorporating Functional Habitats into Conservation Planning for Highly Mobile Species in Dynamic Systems. *Conservation Biology*, Vol. 31, 2017, pp. 1018–1028. http://dx.doi.org/10.1111/cobi.12899.
5. Barclay, J. H., N. M. Korfanta, and M. J. Kauffman. Long-Term Population Dynamics of a Managed Burrowing Owl Colony. *The Journal of Wildlife Management*, Vol. 75, No. 6, 2011, pp. 1295–1306. http://dx.doi.org/doi:10.1002/jwmg.170.

6. White, G., and R. Garrott. *Analysis of Wildlife Relocation Data*. Academic Press, New York City, NY, 1990.
7. McCracken, G. F., K. Safi, T. H. Kunz, D. K. Dechmann, S. M. Swartz, and M. Wikelski. Airplane Tracking Documents the Fastest Flight Speeds Recorded for Bats. *Royal Society Open Science*, Vol. 3, No. 11, 2016, p. 160398. http://dx.doi.org/doi:10.1098/rsos.160398.
8. DeVault, T. L., J. L. Belant, B. F. Blackwell, and T. W. Seamans. Interspecific Variation in Wildlife Hazards to Aircraft: Implications for Airport Wildlife Management. *Wildlife Society Bulletin*, Vol. 35, No. 4, 2011, pp. 394–402. http://dx.doi.org/10.1002/wsb.75.
9. Amelon, S. K., F. R. Thompson, III, and J. J. Millspaugh. Resource Utilization by Foraging Eastern Red Bats (*Lasiurus borealis*) in the Ozark Region of Missouri. *Journal of Wildlife Management*, Vol. 78, No. 3, 2014, pp. 483–493. http://dx.doi.org/10.1002/jwmg.685.
10. Jachowski, D. S., J. B. Johnson, C. A. Dobony, J. W. Edwards, and W. M. Ford. Space Use and Resource Selection by Foraging Indiana Bats at the Northern Edge of their Distribution. *Endangered Species Research*, Vol. 24, No. 2, 2014, pp. 149–157. https://dx.doi.org/10.3354/esr00594.
11. Bergeson, S. M., T. C. Carter, and M. D. Whitby. Partitioning of Foraging Resources between Sympatric Indiana and Little Brown Bats. *Journal of Mammalogy*, Vol. 94, No. 6, 2013, pp. 1311–1320. https://dx.doi.org/10.1644/12-MAMM-A-311.
12. Sparks, D. W., C. M. Ritzi, J. E. Duchamp, and J. O. Whitaker. Foraging Habitat of the Indiana Bat (*Myotis sodalis*) at an Urban-Rural Interface. *Journal of Mammalogy*, Vol. 86, No. 4, 2005, pp. 713–718. https://dx.doi.org/10.1644/1545-1542(2005)086[0713:FHOTIB]2.0.CO;2.
13. Womack, K. M., S. K. Amelon, and F. R. Thompson. Resource Selection by Indiana Bats during the Maternity Season. *Journal of Wildlife Management*, Vol. 77, No. 4, 2013, pp. 707–715. https://dx.doi.org/10.1002/jwmg.498.
14. Miller, D. A., E. B. Arnett, and M. J. Lacki. Habitat Management for Forest-Roosting Bats of North America: A Critical Review of Habitat Studies. *Wildlife Society Bulletin*, Vol. 31, No. 1, 2003, pp. 30–44.
15. Whitaker, J. O. Jr. *Report of Findings: 1997 Field Studies for Interim Mitigation Indiana Bat (Myotis sodalis) at Indianapolis International Airport Marion and Hendricks Counties, Indiana*. U.S. Fish & Wildlife Service, Bloomington, IN, 1998, p. 41.
16. Whitaker, J. O. Jr., D. W. Sparks, and V. Brack, Jr. Use of Artificial Roost Structures by Bats at the Indianapolis International Airport. *Environmental Management*, Vol. 38, No. 1, 2006, pp. 28–36. https://dx.doi.org/10.1007/s00267-005-0117-2.
17. Brigham, R. M. Flexibility in Foraging and Roosting Behaviour by the Big Brown Bat (*Eptesicus fuscus*). *Canadian Journal of Zoology*, Vol. 69, No. 1, 1991, pp. 117–121. http://dx.doi.org/10.1139/z91-017.
18. U.S. Fish and Wildlife Service. *Indiana Bat (Myotis sodalis) Draft Recovery Plan: First Revision*. U.S. Fish and Wildlife Service, Fort Snelling, MN, 2007, p. 258.
19. Sparks, D. W., V. Brack, Jr., J. O. Whitaker, Jr., R. Lotspeich, P. B. Larauge, and M. E. Castille. Reconciliation Ecology and the Indiana Bat at Indianapolis International Airport. In *Airports: Performance, Risks, and Problems*. Nova Science Publishers, Inc, Hauppauge, New York City, NY, 2009, pp. 51–65.
20. U.S. Fish & Wildlife Service. *Biological Opinion on the Application for an Incidental Take Permit for the Federally Endangered Indiana Bat (Myotis sodalis) for the Six Points Road Interchange and Associated Development, Hendricks and Marion Counties, Indiana*. U.S. Fish & Wildlife Service, Bloomington, IN, 2002, p. 36.
21. Whitaker, J. O. S. Jr., D. W. Sparks, J. G. Boyles, D. Judy, J. L. Burskey, and C. M. Ritzi. *2002 Monitoring Program for the Indiana Myotis (Myotis sodalis) Near the Future Six Points Interchange in Hendricks and Marion Counties, Indiana as Required Under the Six Points Interchange Habitat Conservation Plan*. U.S. Fish & Wildlife Service, Bloomington, IN, 2002, p. 100.
22. Sallee, K. L., B. Sallee-Kereszturi, and K. J. Sernka. *Location of a Signal (LOAS) v4.0*. Ecological Software Solutions, LLC, Hegymagas, Hungary, 2010.
23. Horne, J. S., and E. O. Garton. Likelihood Cross-Validation Versus Least Squares Cross-Validation for Choosing the Smoothing Parameter in Kernel Home-Range Analysis. *Journal of Wildlife Management*, Vol. 70, No. 3, 2006, pp. 641–648. http://dx.doi.org/10.2193/0022-541X(2006)70[641:LCVLSC]2.0.CO;2.
24. Millspaugh, J. J., R. M. Nielson, L. McDonald, J. M. Marzluff, R. A. Gitzen, C. D. Rittenhouse, M. W. Hubbard, and S. L. Sheriff. Analysis of Resource Selection Using Utilization Distributions. *Journal of Wildlife Management*, Vol. 70, No. 2, 2006, pp. 384–394. http://dx.doi.org/10.2193/0022-541X(2006)70[384:AORSUU]2.0.CO;2.
25. Aebischer, N. J., P. A. Robertson, and R. E. Kenward. Compositional Analysis of Habitat Use from Animal Radio-Tracking Data. *Ecology*, Vol. 74, No. 5, 1993, pp. 1313–1325. http://dx.doi.org/10.2307/1940062.
26. Bates, D. M., M. Mächler, B. Bolker, and S. Walker. Fitting Linear Mixed-Effects Models Using lme4. *Journal of Statistical Software*, Vol. 67, No. 1, 2015, pp. 1–48.
27. R Core Team. *R: A Language and Environment for Statistical Computing*. R Foundation for Statistical Computing, Vienna, Austria, 2017. http://www.R-project.org/.
28. Burnham, K. P., and D. R. Anderson. Multimodel Inference: Understanding AIC and BIC in Model Selection. *Sociological Methods & Research*, Vol. 33, No. 2, 2004, pp. 261–304. http://dx.doi.org/10.1177/0049124104268644.
29. Barclay, R. M. R. The Effect of Reproductive Condition on the Foraging Behavior of Female Hoary Bats, *Lasiurus cinereus*. *Behavioral Ecology and Sociobiology*, Vol. 24, No. 1, 1989, pp. 31–37. http://dx.doi.org/10.1007/BF00300115.
30. Audet, D. Foraging Behavior and Habitat Use by a Gleaning Bat, *Myotis myotis* (Chiroptera: Vespertilionidae). *Journal of Mammalogy*, Vol. 71, No. 3, 1990, pp. 420–427. http://dx.doi.org/10.2307/1381955.
31. Menzel, J. M., W. M. Ford, M. A. Menzel, T. C. Carter, J. E. Gardner, J. D. Garner, and J. E. Hofmann. Summer

Habitat Use and Home-Range Analysis of the Endangered Indiana Bat. *Journal of Wildlife Management*, Vol. 69, No. 1, 2005, pp. 430–436. http://dx.doi.org/10.2193/0022-541X(2005)069<0430:SHUAHA>2.0.CO;2.

32. Zurcher, A. A., D. W. Sparks, and V. J. Bennett. Why the Bat Did Not Cross the Road? *Acta Chiropterologica*, Vol. 12, No. 2, 2010, pp. 337–340. http://dx.doi.org/10.3161/150811010X537918.
33. Whitaker, J. O. Prey Selection in a Temperate Zone Insectivorous Bat Community. *Journal of Mammalogy*, Vol. 85, No. 3, 2004, pp. 460–469. http://dx.doi.org/10.1644/1383943.
34. Feldhamer, G. A., T. C. Carter, and J. O. Whitaker, Jr. Prey Consumed by Eight Species of Insectivorous Bats from Southern Illinois. *American Midland Naturalist*, Vol. 162, No. 1, 2009, pp. 43–51. http://dx.doi.org/10.1674/0003-0031-162.1.43.
35. O'Keefe, J. M., S. M. Bergeson, L. K. Castor, T. J. Divoll, J. P. S. Hoeh, and B. L. Walters. *2016 Monitoring Program for the Indiana Bat (Myotis sodalis) Near the Six Points Interchange in Hendricks and Marian Counties, Indiana as Required Under the Six Points Interchange Habitat Conservation Plan*. U.S. Fish & Wildlife Service Report, Bloomington, IN, 2017, p. 78.
36. Baxter, D. J. M., J. M. Psyllakis, M. P. Gillingham, and E. L. O'Brien. Behavioural Response of Bats to Perceived Predation Risk While Foraging. *Ethology*, Vol. 112, No. 10, 2006, pp. 977–983. http://dx.doi.org/10.1111/j.1439-0310.2006.01249.x.
37. Scanlon, A. T., and S. Petit. Effects of Site, Time, Weather and Light on Urban Bat Activity and Richness: Considerations for Survey Effort. *Wildlife Research*, Vol. 35, No. 8, 2009, pp. 821–834. http://dx.doi.org/10.1071/WR08035.
38. Cravens, Z. M., V. A. Brown, T. J. Divoll, and J. G. Boyles. Illuminating Prey Selection in an Insectivorous Bat Community, Exposed to Artificial Light at Night. *Journal of Applied Ecology*, Vol. 55, 2017, pp. 705. http://dx.doi.org/10.1111/1365-2664.13036.
39. Russo, D., L. Cistrone, N. Libralato, C. Korine, G. Jones, and L. Ancillotto. Adverse Effects of Artificial Illumination on Bat Drinking Activity. *Animal Conservation*, Vol. 20, No. 6, 2017, pp. 492–501. http://dx.doi.org/10.1111/acv.12340.
40. Pitcock, T. M. *Kennedy Space Center Florida Scrub-Jay Compensation Plan*. NASA Technical Report KSC-E-DAA-TN13444. Kennedy Space Center, FL, 2014, p. 66.
41. Moore, M. D. Personal Air Vehicles: A Rural/Regional and Intra-Urban On-Demand Transportation System. *Journal of the American Institute of Aeronautics and Astronautics (AIAA)*, No. 2003-2646, 2003, p. 20.
42. Langwig, K. E., W. F. Frick, J. T. Bried, A. C. Hicks, T. H. Kunz, and A. M. Kilpatrick. Sociality, Density-Dependence and Microclimates Determine the Persistence of Populations Suffering from a Novel Fungal Disease, White-Nose Syndrome. *Ecology Letters*, Vol. 15, No. 9, 2012, pp. 1050–1057. http://dx.doi.org/10.1111/j.1461-0248.2012.01829.x.
43. Kunz, T. H., and J. D. Reichard. Status Review of the Little Brown Myotis (*Myotis lucifugus*) and Determination that Immediate Listing under the Endangered Species Act is Scientifically and Legally Warranted. Boston University's Center for Ecology and Conservation Biology, Boston, Mass., 2010, p. 30. http://www.bu.edu/cecb/files/2010/12/Final-Status-Review.pdf.
44. Farrow, L. J., and H. G. Broders. Loss of Forest Cover Impacts the Distribution of the Forest-Dwelling Tri-Colored Bat (*Perimyotis subflavus*). *Mammalian Biology*, Vol. 76, No. 2, 2011, pp. 172–179. http://dx.doi.org/10.1016/j.mambio.2010.04.004.
45. Fenton, M. B., and R. M. R. Barclay. *Myotis lucifugus. Mammalian Species*, Vol. 142, 1980, pp. 1–8. http://dx.doi.org/10.2307/3503792.
46. Fujita, M. S., and T. H. Kunz. *Pipistrellus subflavus. Mammalian Species*, Vol. 228, 1984, pp. 1–6. https://doi.org/10.2307/3504021.
47. Nelson, J. J., and E. H. Gillam. Selection of Foraging Habitat by Female Little Brown Bats (*Myotis lucifugus*). *Journal of Mammalogy*, Vol. 98, No. 1, 2017, pp. 222–231. http://dx.doi.org/10.1093/jmammal/gyw181.
48. Bell, D., J. Hjältén, C. Nilsson, D. Jørgensen, and T. Johansson. Forest Restoration to Attract a Putative Umbrella Species, the White-Backed Woodpecker, Benefited Saproxylic Beetles. *Ecosphere*, Vol. 6, No. 12, 2015, pp. 1–14. http://dx.doi.org/10.1890/ES14-00551.1.

The ACRP Selection Panel for the Graduate Research Award Program on Public-Sector Aviation Issues peer-reviewed this paper (18-01543).

TRR
JOURNAL OF THE TRANSPORTATION RESEARCH BOARD

Article

Transportation Research Record
2018, Vol. 2672(29) 23–34

DOI: 10.1177/0361198118780681
journals.sagepub.com/home/trr
SAGE

Developing Machine-Learning Models to Predict Airfield Pavement Responses

Osman Erman Gungor[1] and Imad L. Al-Qadi[1]

Abstract

Aviation promotes trade and tourism by connecting regions, people, and countries. Having a functional and efficient airport pavement network is important to improve aviation traffic and to provide safer mobility to almost 800 million passengers travelling in the U.S. per year. The Federal Aviation Administration has initiated and actively been participating in many projects to further advance pavement design and performance to meet user requirements. To accomplish that, quantitative data are needed; such data may be collected from the pavement response to gear and environment loading. In this study, responses from four instrumented taxiway concrete slabs at John F. Kennedy International Airport were analyzed. The collected data were used to develop machine-learning (ML) based prediction models to compute the temperature, curling and bending strains within pavement. The ML models were developed using the support vector machine (SVM) algorithm. The results showed that SVM based ML models can predict pavement responses with a high accuracy and low computation time. Furthermore, in the case of feeding more data from various airports, ML models have proven to be a promising technique for pavement analysis engine for future airport pavement design frameworks. This study also produces recommendations for future data collection projects to have well-designed databases for data-driven models development.

Aviation promotes trade and tourism by connecting regions, people and countries. According to IATA (*1*), in 2015, aviation transported approximately 3.6 billion passengers and carried 51.2 tons of freight, which composes 35% of interregional exports and goods by value. Moreover, these transportation activities created 9.9 million jobs worldwide and accounted for 2.7 trillion dollars of the total global economy.

Having a functional and efficient airport pavement network is important to support mobility and freight demands. The U.S. has approximately 9,000 paved airports with approximately 650 million square yards of paved runway surface. The conditions of these pavements are critical to provide safe mobility to almost 800 million passengers travelling in the U.S. per year. Therefore, to sustain a well-maintained and functional airport pavement system, the Federal Aviation Administration (FAA) spends about $4 billion per year for rehabilitating and repairing airport pavements (*2*).

One of the efficient ways to reduce rehabilitation and repairing costs is to better understand airport pavement performance and response to gear and environmental loading. For this purpose, FAA has started instrumenting airport pavements with pressure, strain and temperature sensors to monitor pavement behavior under aircraft and environmental loads to allow for more accurate analysis and design of airport pavements.

FAA uses a state-of-the-art mechanistic–empirical design approach for rigid pavement (*3*). In this approach, the structural responses within the pavement are computed using finite element (FE) analysis. The calculated responses are used to predict potential distresses during the pavement service life. For example, the maximum tensile stress at the bottom of a concrete slab is used to predict the fatigue life of the pavement. FE analysis is superior to conventional analysis techniques (e.g., layered elastic theory or Westergard slab theory) since it can consider non-linearities in loading conditions and material characterization. However, it still carries the inherent limitations of mechanistic approaches: simplifying or neglecting variables in the complicated tire–pavement interaction.

With improvements in data storage and collection, machine learning (ML) is gaining momentum as a

[1]University of Illinois at Urbana-Champaign, Illinois Center for Transportation, Rantoul, IL

Corresponding Author:
Address correspondence to Osman Erman Gungor: gungor2@illinois.edu

potential behavior prediction method in the field of engineering (*4*, *5*). ML may be defined as the development of algorithms that iteratively learn from data and produce reliable, repeatable decisions and results (*6*). Its ability to capture complicated relations in the data makes it a promising candidate to address challenging cases in engineering that would be difficult to solve with the traditional mechanistic approaches. Additionally, unlike the mechanistic approaches, it provides computational efficiency since it does not require rebuilding the model (i.e., training) every time different inputs are introduced. Therefore, this study suggests developing ML based, data-driven models to compute pavement responses with low computational time.

This paper employs support vector machine to develop ML models to compute temperature, curling strain (due to the environmental loading) within the pavement and bending strain at the bottom of concrete slabs under Boeing 777 loading. The spacing of sensors installed allows for measurement of responses for this type of aircraft gear geometry. Although data from Airbus A380 could be captured, the data were limited (only eight data points were available) for model development. Additionally, this paper presents an online application that integrates the ML models. This web-based tool could be used by researchers and engineers to validate the pavement responses from mechanistic models used with various sets of inputs. It can be ultimately used for pavement damage prediction.

Data Description

FAA, in collaboration with Port Authority of New York and New Jersey (PANYNJ), heavily instrumented four concrete slabs of a taxiway at John F. Kennedy International Airport (JFK) in August, 2010 (*7*). The pavement section of the taxiway has 20 in concrete slab supported by two base layers. The first base layer is a 4-inch-thick asphalt-modified base and the second base layer is a 6-inch-thick dense graded granular base.

In total, 40 strain gauges, eight pressure cells, and six thermocouples were installed. The instrumentation layout is given in Figure 1. The strain gauges labelled A were instrumented at the mid-depth of the slab and label B strain gauges were instrumented at the bottom of the slab. Furthermore, the temperature gauges were placed at three different depths: 10, 16, and 19 in. Lastly, pressure cells were embedded at the interface between the concrete slab and the asphalt-modified base.

An additional dataset, meteorological data from JFK airport, was included in the study to predict pavement responses at given weather conditions. The following sections present the data used in this study.

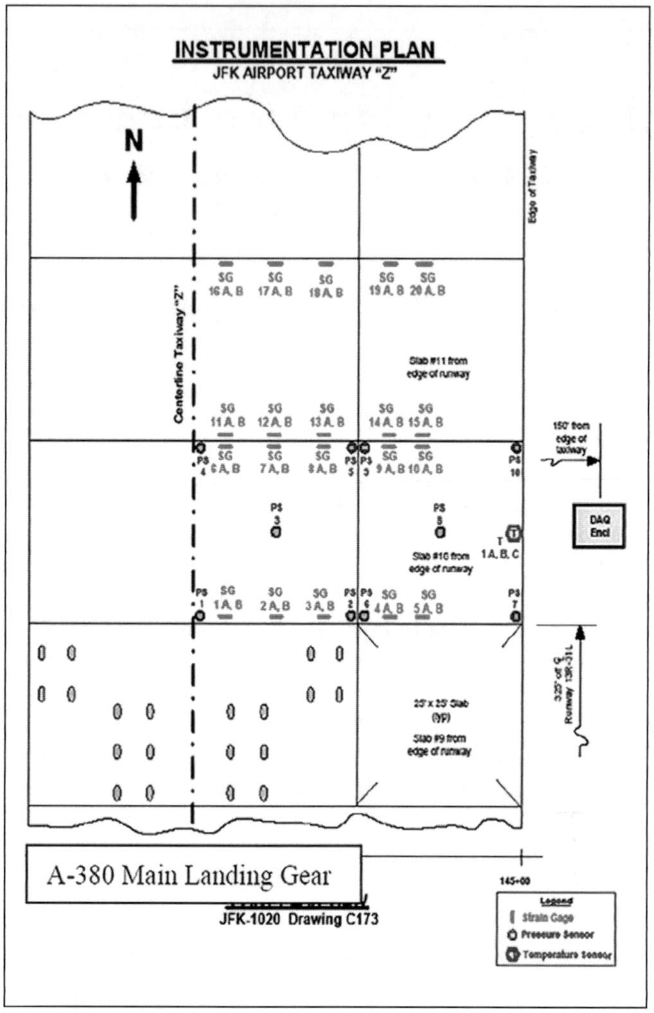

Figure 1. Instrumentation layout (*7*).

Static Data

Static data were collected hourly at the beginning of each hour for 5 seconds at 5 Hz when the pavement is not loaded by aircraft. The purpose of collecting static data was to study the concrete slab curling. The static data were provided in ten different files. Each data file has its own starting and ending day and time. However, data collection was interrupted between starting and ending dates for various technical issues, such as solar energy interruption due to weather conditions. Table 1 summarizes the static data considered showing starting and ending dates and missing sensors/data for each file.

Dynamic Data

The dynamic data were collected under various aircraft loadings. Aircraft identification was based on pictures captured by a camera mounted on the data acquisition system (DAS) cabinet. However, no images were captured at specific times, or the captured images could not

Table 1. Static Data Summary

Initial date	Initial time	Ending date	Ending time	Missing sensors	# Data points	# Missing hours
15-Apr-11	16:00	25-Apr-11	12:00	None	256	5
4-May-11	13:00	27-May-11	07:00	None	269	278
3-Jan-12	13:00	1-Feb-12	18:00	9 strain gauges	542	160
1-Feb-12	20:00	21-Feb-12	19:00	9 strain gauges & 6 temperature sensors	467	13
6-Mar-12	12:00	25-Mar-12	09:00	9 strain gauges	282	172
2-Apr-12	09:00	22-May-12	20:00	9 strain gauges	940	272
13-Jul-12	08:00	31-Jul-12	08:00	9 strain gauges	375	57
13-Aug-12	07:00	11-Sep-12	09:00	9 strain gauges	535	163
5-Dec-12	14:00	2-Jan-13	12:00	9 strain gauges	666	5
31-Jan-13	15:00	3-Feb-13	13:00	9 strain gauges	71	0

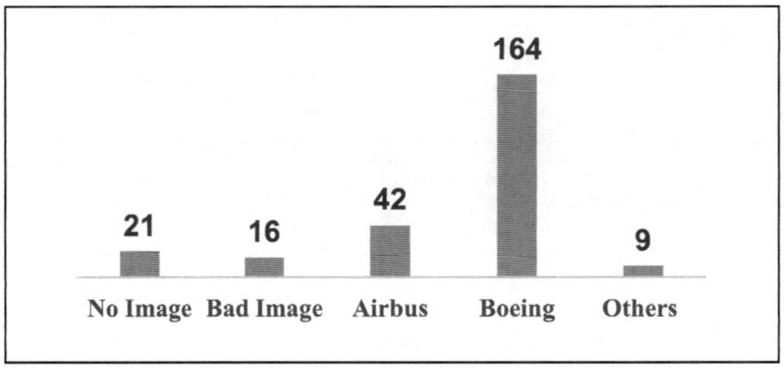

Figure 2. Overall aircraft distribution.

be used to identify the type of aircraft. The overall result of aircraft picture distribution is given in Figure 2.

The aircraft model distribution within Airbus and Boeing is given in Figure 3. The aircraft models in the x-axis are ranked based on their size, ranging from small to large. It should be noted that almost all models have a couple of different versions. In this study, all different versions are grouped under the name of the model. For example, Boeing 777-200 and Boing 777-200ER are grouped under Boeing 777.

Meteorological Data

National Centers for Environmental Information (*8*) is the website published and maintained by the National Oceanic and Atmospheric Administration. This website provides a variety of climatic data from different regions within the U.S. The meteorological data from the JFK station were requested and downloaded from this website. There are 17 important variables in this dataset. The name, type, and explanation for each variable are given in Table 2.

Data Pre-Processing

Pre-processing is an important step in any data analytic study. It basically means converting the raw data into analytic data that can be used for prediction models. The pre-processing for each dataset is explained below.

Static Data

As previously mentioned, the static data were collected hourly at the beginning of each hour for 5 seconds at 5 Hz. In other words, approximately 25 data points were collected at each hour. These data points, as expected, showed very little variance within 5 seconds since there was no loading. Therefore, their mean was calculated and considered at the data point that corresponds to a particular hour. After that, the variation of measurements with respect to time can be analyzed and predicted.

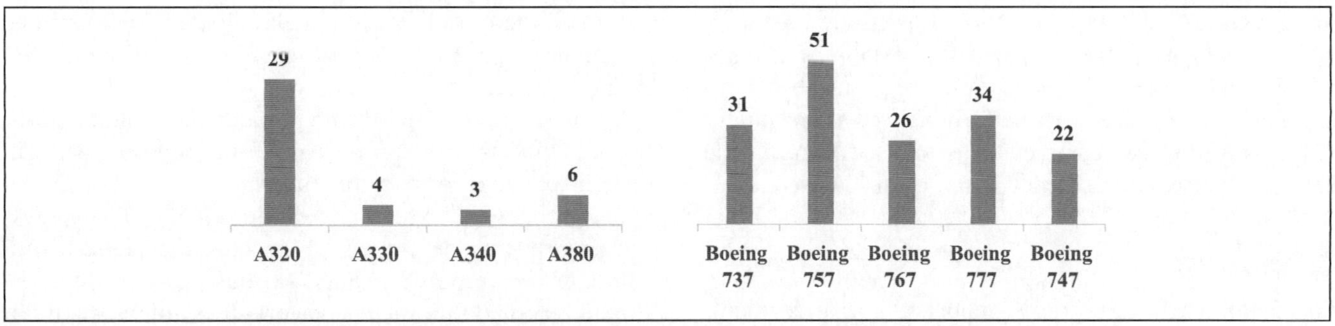

Figure 3. Aircraft model distribution within Airbus and Boeing family.

Table 2. The List of Important Variables in Meteorological Data

Name	Type	Explanation	Unit
Date	Date	The day and time of the data	yy:mm:dd hh:mm:ss
DIR	Continuous	Wind direction	compass degree
SPD	Continuous	Wind speed in miles per hour	miles per hour
GUS	Continuous	Gust: a sudden, brief increase in speed of the wind	miles per hour
CLG	Continuous	Cloud ceiling: lowest opaque layer	hundreds of feet
SKC	Categorical	Sky coverage or cloud coverage	NA
VSB	Continuous	A measure of the distance at which an object or light can be clearly discerned	statute miles to nearest tenth
TEMP	Continuous	Weather temperature	Fahrenheit
DEWP	Continuous	Dew point temperature at which air is saturated with water vapor	Fahrenheit
SLP	Continuous	Sea level pressure	millibars to nearest tenth
ALT	Continuous	Altimeter setting	inches to nearest hundredth
STP	Continuous	Station pressure	millibars to nearest tenth
MAX	Continuous	Maximum temperature	Fahrenheit
MIN	Continuous	Minimum temperature	Fahrenheit
PCP01	Continuous	The precipitation for the preceding 1-hour period	inches and hundredths
PCP06	Continuous	The precipitation for the preceding 6-hour period	inches and hundredths
PCP24	Continuous	The precipitation for the preceding 24-hour period	inches and hundredths
SD	Continuous	Snow depth	inches

Note: NA = not applicable.

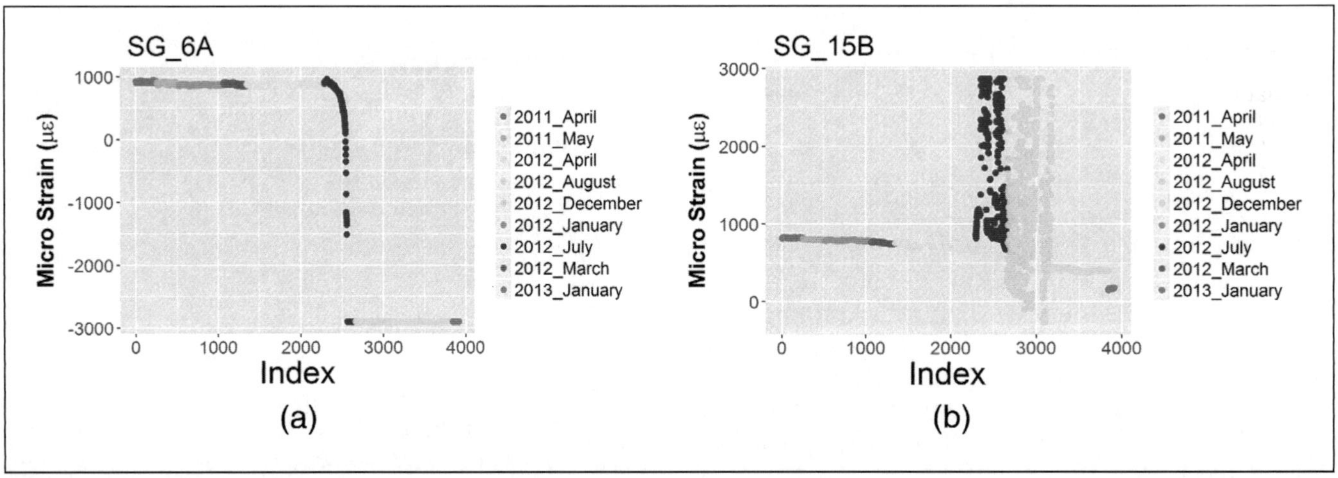

Figure 4. Anomalies for strain gauges (*a*) SG_6A, and (*b*) SG_15B.

Although thermocouples managed to collect data throughout the analysis period, there are some anomalies observed in strain gauge measurements. Figure 4*a* and *b* shows the strain measurements with respect to time for the gauges SG_6A and SG_15B, respectively. As seen, after January 2012, the gauges started exhibiting anomalies. Similarly, other gauges showed the same behavior. Although some gauges kept performing well after January 2012, data after January 2012 was omitted from all the gauges to preserve consistency in size of the datasets.

Dynamic Data

Pavement instrumentation is important for understanding pavement behavior under gear/axle loading. Depending on the approach used for simulating axle or gear loadings, the studies in the literature can be group into three categories: accelerated pavement testing (*9*, *10*), track test (*11*, *12*), and field instrumentation (*13*, *14*). In all these studies, the signals collected from various sensors were processed to extract responses generated by gear or axle loading.

In most studies, speed and wander (i.e., lateral position of vehicles with respect to the sensors) were known or controlled. However, this information was not available for this study which posed additional challenges. A framework was developed to process the signals and compute the responses under various gear loadings. It should be noted that the framework does not only extract peak responses, but also trims the part of the signal that does not carry any information. As a result of this trimming, the size of the signal was reduced, thus leading to a reduction in data storage.

Signal Filtering. Signal filtering for airport pavement responses involves unique challenges. Traditional filtering techniques, such as the moving average, are not ideal for this problem as they may cause a loss of information. When a gear passes over sensors, the signal appears like a spike, which can be deteriorated by the averaging-based filtering technique.

This study employs a technique called "Savitzky-Golay" (*15*). This filtering technique outperforms conventional averaging filtering techniques by preserving the high frequency component of the signals. It essentially fits the user-defined degree of polynomial to the part of the data extracted by a sliding window. The length of the window is also user defined. The degree of polynomial and window length were selected as 2 and 21 seconds, respectively.

Signal Compression. It was observed that the signals may have parts either at the beginning (before gear passes over the sensor) or at the end (after gear leaves the sensor) that do not get affected by loading. These parts may be disregarded to reduce the amount of space required to store the signal. The first step for signal compression is to identify the lower and upper envelopes for each signal (Figure 5*a*). The envelopes are found by fitting spline functions for the signal local maxima and minima. The difference between the two envelopes is calculated at each data point. Starting from both ends of the signal, the data points are removed until the difference between the envelopes is greater than 0.5 microstrain. For this particular signal, approximately 50% reduction was obtained. The compression algorithm was applied to all signals. In average, the signals were compressed by 37.8%.

Sensor Response Identification. The first step for extracting the response of a sensor due to a gear loading is to identify the peaks of the signal. The peaks were identified utilizing a MATLAB function called "findpeaks" (*16*). Spurious peaks in the signal can be eliminated by applying a threshold for the prominence of peaks. The prominence of the peaks can be defined as a "measure of a peak showing how much the peak stands out due to its intrinsic height and its location relative to other peaks" (*17*). Peaks with small prominence could represent noises in the signal and may be removed. The threshold for the prominence was selected as 0.5 microstrain in this study.

The second step is computing a reference point, that is, the strain measurement when the aircraft is far away from the sensor. The reference point is calculated by taking the average of removed points from the previous step. Afterwards, the final response under the gear loading is computed as the difference between the reference point and peaks.

Figure 5*b* demonstrates calculation of the response with respect to the reference points. Trimming Points 1 and 2 is explained under "Signal Compression" above. Later, the reference points are computed by taking the mean of the removed points. If there are two reference points as in Figure 5*b*, the average is assumed to be the final reference point. Finally, the actual response is the distance between the peak point and the final reference point. In Figure 5*b*, two responses (one tension from

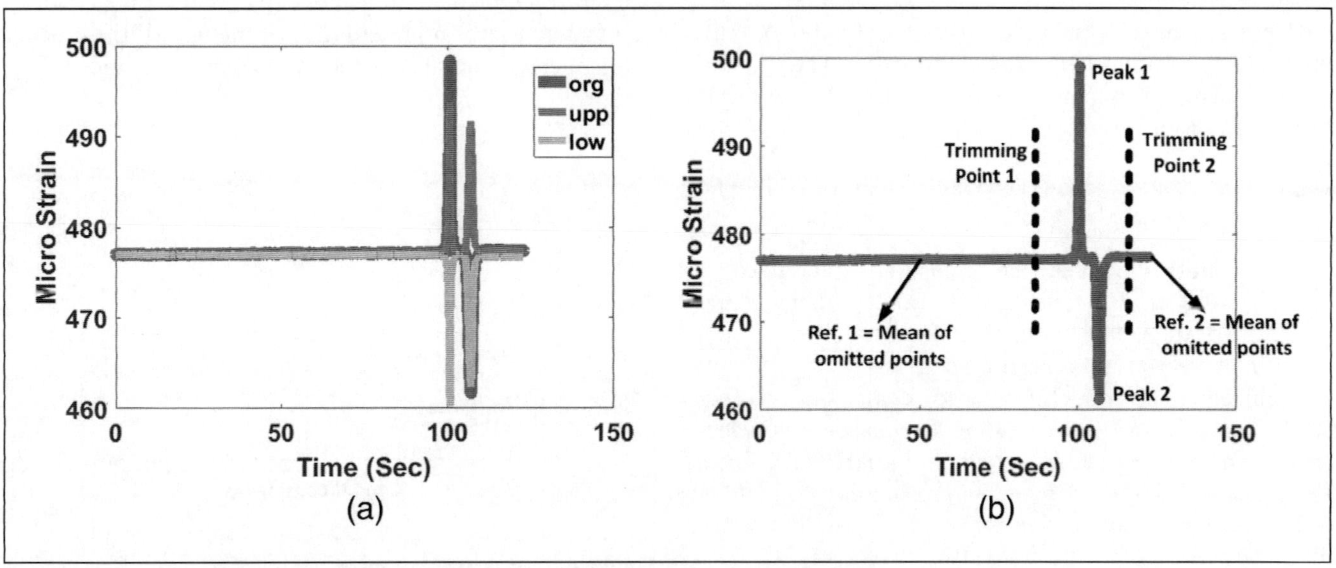

Figure 5. Showing (*a*) lower and upper envelopes for a signal, and (*b*) peak extraction from a signal.

Peak 1 and one compression from Peak 2) were extracted as 21.22 and 15.42 microstrain, respectively. It should be noted that the number of peaks in the signal and type of the responses (e.g., tension or compression) depend on the position of the sensor with respect to the gears and the number of axles in the gears (e.g., single, tandem, or tridem). Instead of having multiple peaks as a result of multi-gears loading, the maximum response was considered to be the response from that aircraft.

Meteorological Data

The meteorological data was downloaded between January, 2011 and February, 2013. The data were collected hourly referencing Greenwich Mean Time Zone. The time was converted into East Time Zone. Later, this dataset was merged with the pavement response data by matching the time in these two datasets.

Some missing data were noted in the meteorological data. In data science, replacing missing data with substituted values is called imputation. Several techniques are available to impute the dataset. This study utilizes a recently developed technique called "MissForest" (*18*). There are two main advantages in using this algorithm. The first one is non-parametric imputation technique, which means that it does not need tuning parameters, or it does not require assumptions about data distribution. The second advantage is its ability to automatically handle mixed-type data, that is, data consisting of both continuous and categorical variables.

Bias–Variance Decomposition for Regression and Stepwise Forward Regression for Variable Selection

In the regression theory, it is assumed that there is an underlying deterministic model (i.e., true model) that represents the output (i.e., variable to be predicted) with only mean-zero random error (i.e., noise). Equation 1 formulates this assumption as

$$Y = f(x) + \varepsilon \tag{1}$$

where

Y= variable to be predicted or dependent variable;

$f(x)$ = underlying deterministic model where x are predictors or independent variable(s); and

ε = mean-zero random error, i.e. $E(\varepsilon) = 0$.

Each regression model developed to predict Y (using ML or not) is an approximation to this deterministic model, which is denoted as a random variable $\tilde{f}(x)$ since it depends on the type of regression technique used.

The objective in regression is to find the $\tilde{f}(x)$ that minimizes the mean square error (MSE):

$$\min_{\tilde{f}(x))} E\left[Y - \tilde{f}(x)\right]^2 \tag{2}$$

MSE can be decomposed using basic mathematical operations and properties of variance and expectation as

$$E\left[Y - \tilde{f}(x)\right]^2 = \left(f(x) - E\left[\tilde{f}(x)\right]\right)^2 + var\left(\tilde{f}(x)\right) + E\left[\varepsilon^2\right] \tag{3}$$

In Equation 3, the first term ($\left(f(x) - E\left[\tilde{f}(x)\right]\right)^2$) and the second term ($var\left(\tilde{f}(x)\right)$) are called bias and variance of a model, respectively. The last term only depends on the noise; therefore, it is called incompressible error. The total error of a prediction model is equal to the summation of these three terms.

As seen, in the case of choosing the simplest model for the prediction function (i.e., $\tilde{f}(x)$ = some constant), the variance of the model becomes zero (variance of a constant is always zero). In contrast, the bias term will be very high because a constant would not be enough to approximate complicated underlying models ($f(x)$). However, when using the second simplest model (i.e., $\tilde{f}(x)$ = ax + b, a linear model), the variance increases (resulting in a non-zero variance because of x) and the bias decreases. This adverse relation between bias and variance is called the bias–variance trade-off; it is represented in Figure 6. Figure 6 shows that an optimum set of variables should be found to minimize the total error of the model. Moreover, it demonstrates that in the case of adding all variables to the model, the bias will be minimized. However, the variance will reach its maximum, thus maximizing the total error of the model.

In the case of using a complicated prediction model (i.e., $\tilde{f}(x)$), it is quite difficult and sometimes impossible to compute the bias and variance terms analytically. Therefore, approximations are used. While the accuracy in a training set is representative of the bias of the model, testing accuracy indicates the total error of the model.

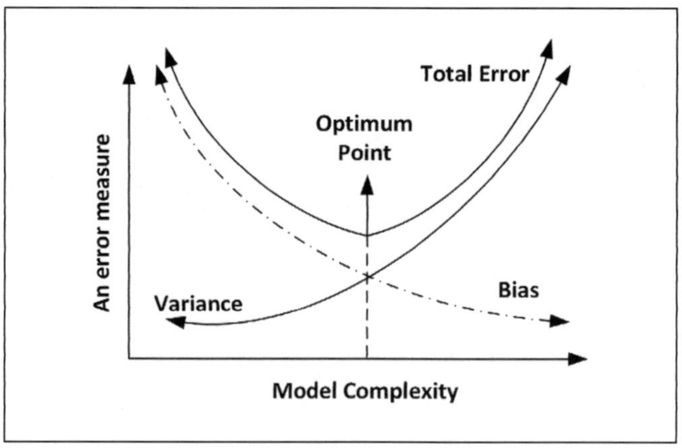

Figure 6. Bias–variance trade-off.

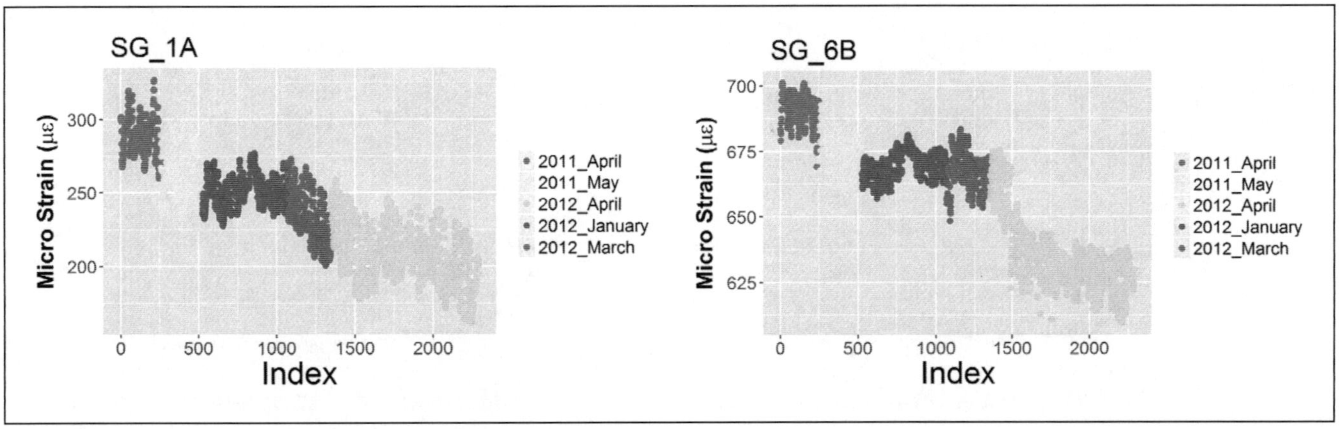

Figure 7. Demonstration of time effect on strain measurements.

Several techniques are suggested in the literature for selecting an optimum set of variables to minimize the total error of a regression model. This paper utilizes the stepwise forward regression technique, which starts finding the best one-variable model that produces the highest testing accuracy. Afterwards, the number of variables in the model keep increasing until a reduction in testing accuracy is observed, that is, the optimum model complexity is reached. Conventionally, the linear regression is used in the stepwise regression. However, in this paper, the authors use the support vector machine with linear kernel (introduced later in the paper). The results of variable selection using stepwise regression are given in the following sections.

Temperature Prediction

Figure 8*a* shows the results of stepwise regression for temperature prediction. It can be clearly seen that the results conform to previously explained bias–variance trade-off (Figure 6). Whereas training accuracy always increases as the number of variable increases (i.e., MSE is decreasing), testing accuracy starts to decrease for the models that have more than seven variables. The optimum variable set was found to be the air temperature, direction and speed of wind, dew point temperature, station pressure, cloud ceiling, and maximum temperature.

The mechanistic approach used to predict temperature within pavement is called "Enhanced Integrated Climatic Model (EICM) (*19*)". The developed ML model shares six of its seven inputs with EICM. However, the station pressure that is found to be a significant variable for temperature prediction is not considered by EICM.

Curling Strain Prediction

The results of variable selection for strain prediction is shown in Figure 8*b*. The variable set that minimizes the total error was found to be air temperature, dew point temperature, station pressure, pavement temperature read by three thermocouples, and age of concrete.

Age of concrete was the product of feature engineering that can be defined as the process of using domain knowledge of the date to create features (variables) to improve the performance of ML algorithms. The motivation behind adding this variable is demonstrated in Figure 7. As seen, there is a clear pattern in strain measurements with respect to time. Although this trend can be attributed to the creep effect of concrete slab, investigating the true reason behind it is left for future studies. It should be noted that age of the concrete is not considered by conventional pavement concrete design and analysis approaches although it turned out to be a significant variable in ML prediction models.

Bending Strain Prediction

Figure 8*c* shows stepwise regression results for bending strain prediction. The identified optimum variable set includes air temperature, dew point temperature, pavement temperatures at 10 in. and 16 in., and curling strain at the bottom of the slab. The curling strain at the bottom of the slab and the temperature within pavement are computed using the prediction models in this study.

Prediction Model Development

Support Vector Machine

This study aims to develop a data-driven pavement analysis engine that inputs the aforementioned, selected explanatory variables and outputs temperature and strains. To develop such an engine, this study employs a robust prediction technique called support vector machine (SVM).

SVM is a supervised ML algorithm developed by Cortes and Vapnik (*20*). SVM adopts the structural risk minimization (SRM) principle that targets to minimize

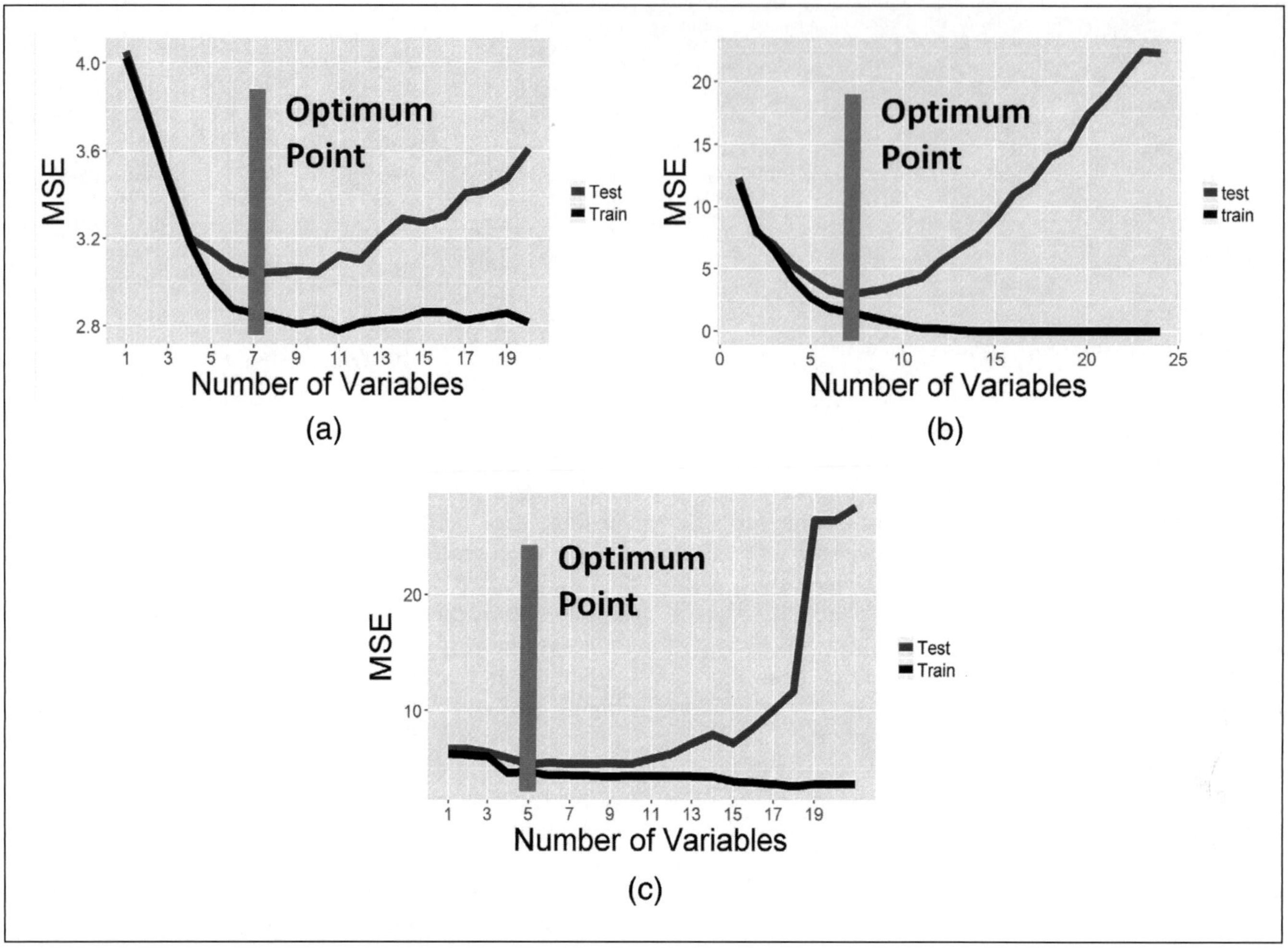

Figure 8. Variable selection results for (*a*) temperature, (*b*) curling strain, and (*c*) bending strain.

the upper bound on the expected error. This is fundamentally different from the empirical risk minimization (ERM) principle that is employed by conventional neural networks (*21*). ERM is based on minimizing the error on the training data which may lead to overfitting problems in the case of having smaller datasets. SVM implements SRM as a constrained optimization problem given in Equation 4 as

$$\frac{1}{2} \| w\|^2 + C\sum_{i=1}^{N} \xi_i \quad (4)$$

$$y_i\left(w^T x_i + b\right) \geq 1 - \xi_i, \xi_i \geq 0, i = 1 \ldots N \quad (5)$$

where

w and b = model parameters;
C = tuning parameter for margin and training error;
ξ_i = slack variable for defining soft margin; and
N = the number of data points.

SVM is capable of separating data without error if the data are linearly separable. However, most of the engineering problems including predicting the pavement responses are inherently non-linear. SVM tackles the problem with non-linearities by mapping the training data into a higher dimension space where the data is hoped to be more linear (Figure 9). There are many kernel functions developed in the literature. The most

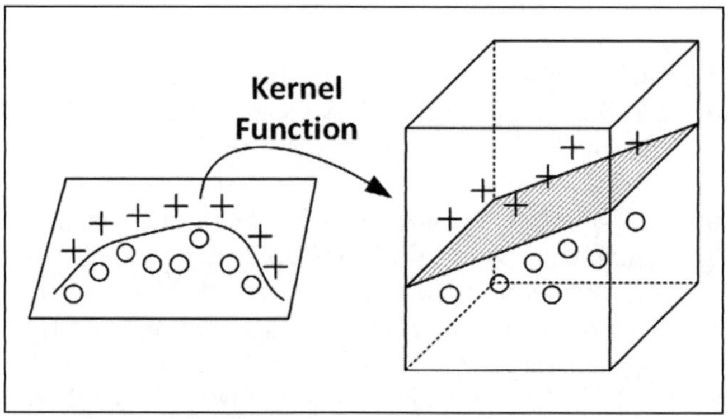

Figure 9. Kernel trick.

Table 3. Results for Temperature Predictions

Temperature			Training		Testing	
Sensor	Depth (inches)	Statistics	R^2	MSE (°F)	R^2	MSE (°F)
T_1A	10	Mean	0.983	0.917	0.953	2.295
		Std	0.001	0.014	0.005	0.086
T_1B	15	Mean	0.979	0.966	0.945	2.390
		Std	0.001	0.017	0.003	0.059
T_1C	19	Mean	0.976	0.986	0.939	2.404
		Std	0.000	0.016	0.006	0.111
T_2A	10	Mean	0.984	0.874	0.955	2.276
		Std	0.000	0.002	0.004	0.102
T_2B	15	Mean	0.979	0.972	0.945	2.374
		Std	0.001	0.015	0.006	0.100
T_2C	19	Mean	0.977	0.970	0.941	2.373
		Std	0.001	0.019	0.007	0.115

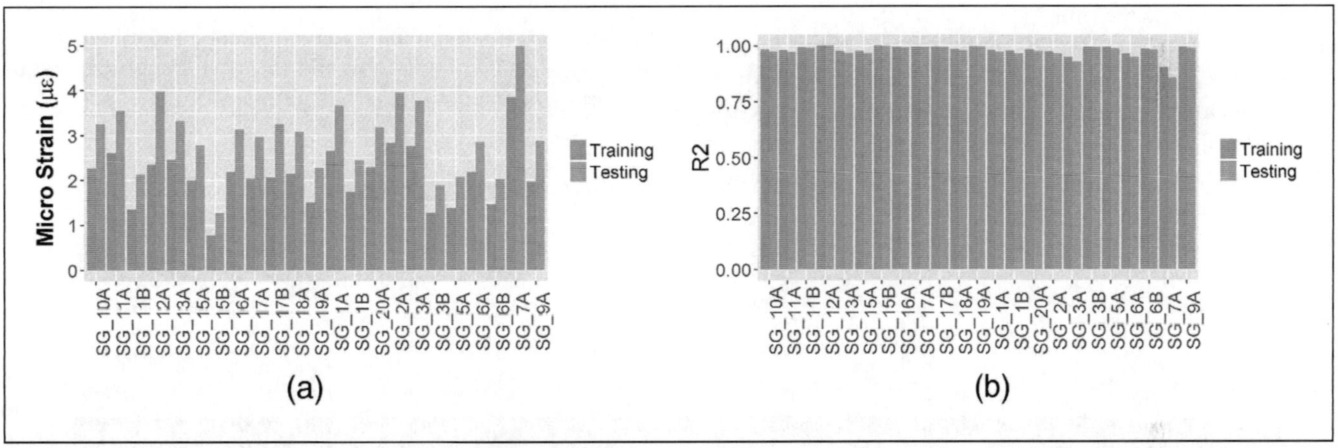

Figure 10. Strain prediction results in terms of (*a*) MSE, and (*b*) R^2.

common kernel types are radial-based kernel, polynomial kernel, Gaussian kernel, and exponential kernels. The accuracy of SVM models is highly affected by the type of the kernel (*22*). Moreover, performance of kernel functions is problem dependent and the right kernel type should be determined by conducting validation step during the training of the models. In this study, the radial-based kernel (Equation 6) was found to be the most efficient kernel type for computing the pavement responses.

$$e^{\ \gamma\|x\ \ x'\|^2} \qquad (6)$$

where

γ = scaling parameter, and
x and x' = data samples.

Model Results

The datasets were randomly divided into three parts: training (80%), validation (10%), and testing dataset (10%). The training dataset was used to compute SVM model parameters. Additionally, similar with the majority of machine learning algorithms, SVM has also the so-called hyper parameters that are essentially user inputs such as kernel type, cost coefficient and so on. Cross validation step was conducted to select the hyper parameters that maximized the accuracy of trained model in validation dataset. Finally, accuracy of the trained model with tuned hyper parameters are evaluated by testing dataset.

Table 3 shows accuracy results for temperature prediction. As previously mentioned, the data was randomly split into various parts. To eliminate the effect of randomness, the splitting was done 10 times. Therefore, the results are presented with statistical measure. As seen, MSE of the models for almost all the sensors is around 2.4°F with R^2 value of 0.97.

Figure 10*a* and *b* shows the results for curling strain prediction for various sensors (average of 10 runs) in terms of MSE and R^2, respectively. The labels of the sensors are given in the *x*-axis. As seen, MSE for testing

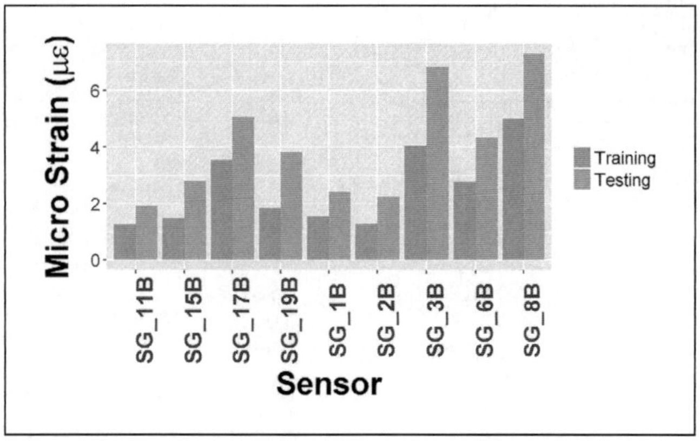

Figure 11. Prediction results.

dataset is around a couple of micro strain and the R^2 values are around 0.98. The strain gauges that failed after instrumentation were omitted.

As previously mentioned, the ML models for Boeing 777 were developed based on 35 data points. An important implication associated with a small dataset is the sensitivity of the model performance to data splitting. In other words, performance of the models can significantly change depending on which data points the models are trained and tested on. To overcome this problem and report the true performance of models, random splitting was repeated 1,000 times.

Figure 11 presents the MSE of the prediction models (average of 1,000 runs) for Boing 777 loading and shows that the SVM-based ML models could predict the responses with MSE range of 1.5 to 7 microstrain. The strain gauges placed at the mid-depth of the slab were not considered in the prediction models, because they are close to neutral axis of the slab where the bending strains approach zero. Furthermore, the strain gauges that failed after instrumentation were omitted.

There are two important variables missing in the dataset that might hinder the performance of the models. The first one is traffic wander, which is the lateral position of the gear with respect to sensor locations. The significance of wander on pavement responses has been highlighted by several studies (*23*, *24*). The second one is speed. Although aircraft speed on taxiway is expected to have a marginal effect on pavement responses, such information could be useful when extracting the peak response from the signal. Incorporating these two variables into data collection would improve the efficiency of ML models.

Online Web Application

An online application that integrates the developed ML models for both static and dynamic data (Figure 12) was developed. In this application, the user can construct one

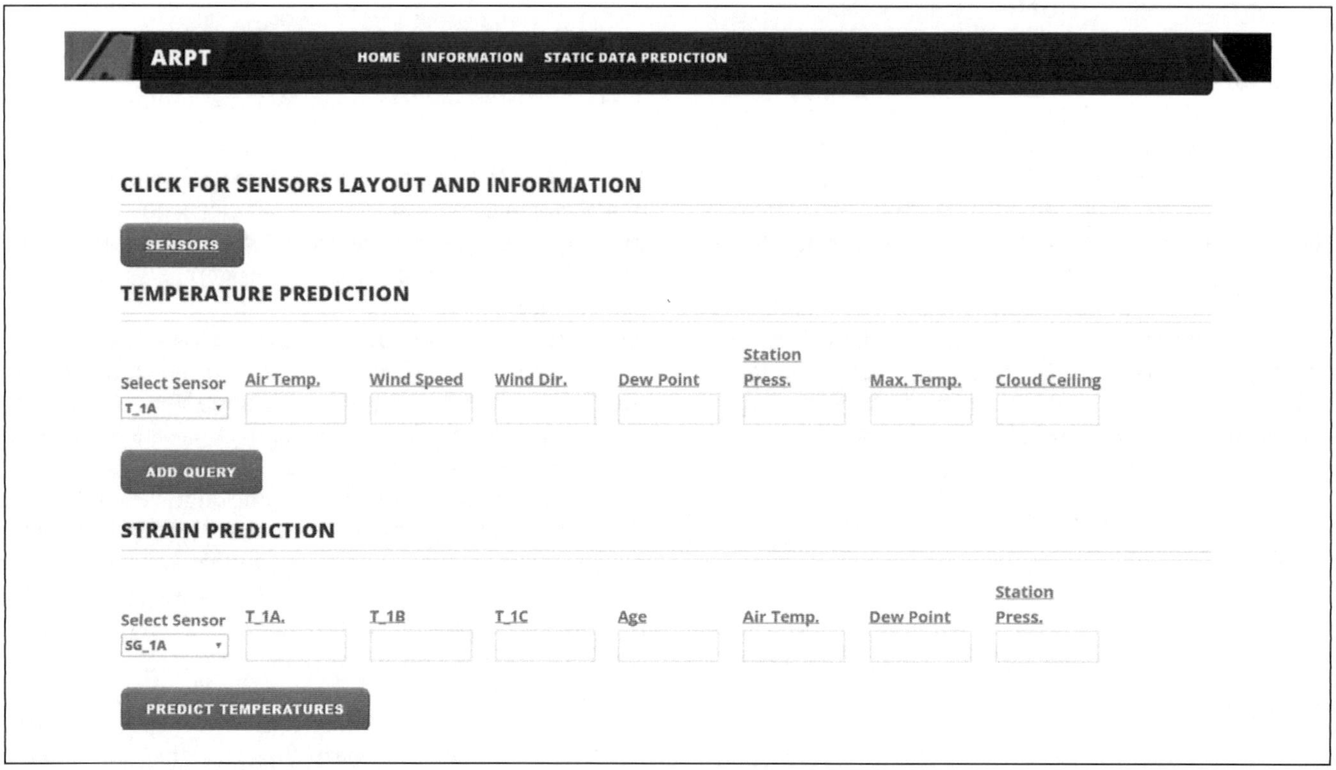

Figure 12. A snapshot from the online application.

or more queries depending on which response computation is desired. There is no limitation to the number of queries that can be submitted. The average computation time per query is 0.16 sec, which is significantly smaller when compared with the computational time of mechanistic models. This web-based tool could be used by researchers and engineers to validate the pavement responses from mechanistic models used with various sets of inputs. It can be ultimately used for pavement damage prediction.

Summary and Conclusions

A well-maintained and functional airport pavement network is essential to sustain people and goods mobility. The FAA has been proactively involved in many projects to further advance pavement design, analysis, and rehabilitation procedures. Four taxiway concrete slabs at JFK airport were instrumented using various sensors to better understand pavement behavior in response to airplane gear and environmental loadings. This study uses the collected data from this taxiway to develop ML based prediction models to compute temperature, curling and bending strains within the concrete pavement.

Current mechanistic pavement analysis approaches utilize assumptions that overlook pavement–tire interaction. This study showed that ML based data-driven models are a promising candidate since they simulate the pavement under field conditions and are able to predict responses with quite good accuracies. This study showed that ML model has the potential to eventually be used as pavement analysis engine in future pavement design frameworks. Although the results presented herein are specific to one pavement section, these models can be extended by integrating data from other airports.

Although the results show that ML models could predict pavement responses under Boeing 777 with reasonable accuracy (MSE ranging between 1.5 to 7 microstrain), the models could be further improved by adding two more variables into the database: wander and speed. Especially, the lateral position of aircraft with respect to the location of sensors (i.e., wander) is expected to have a significant impact on the accuracy of the models.

Additionally, it is recommended to build a design of experiment (DOE) method into the process of dynamic data collection to systematically determine the value of the parameters (e.g., the number of aircraft for each model and temperature) so that the collected data could be efficiently used for developing more generalized prediction models. Furthermore, using a DOE method would help collect data to quantitatively explain the relation between output and particular parameter in the experiment.

Acknowledgments

This work was sponsored by the Airport Cooperative Research Program through a Graduate Research Award. The authors acknowledge Larry Goldstein, Navneet Garg, David Brill, Berg Randall and Sarah Pauls for their guidance and investment in this research.

References

1. International Air Transport Association. *Industry Statistics. Fact Sheets*. http://www.iata.org/pressroom/facts_figures/fact_sheets/Pages/index.aspx. Accessed June 10, 2017.
2. Agrawal, S. *Paving the Way to Aviation's Future*. National Airport Pavement Test Facility 10th Anniversary Celebration, FAA, July 23, 2009. http://www.airtech.tc.faa.gov/NAPTF10Years/PaveFuture.asp.
3. Federal Aviation Administration. *Series 150 Advisory Circulars (ACs) for Airport Projects*, https://www.faa.gov/airports/resources/advisory_circulars/
4. Nabian, M. A., and H. Meidani. Uncertainty Quantification and PCA-Based Model Reduction for Parallel Monte Carlo Analysis of Infrastructure System Reliability. Presented at 96th Annual Meeting of the Transportation Research Board, Washington D.C., 2017.
5. Nabian, M. A., and H. Meidani. Deep Learning for Accelerated Reliability Analysis of Infrastructure Networks. Computer-Aided Civil and Infrastructure Engineering, Vol. 33, 2018, pp. 443–458.
6. Sathiyamoorthi, V. Data Mining and Data Warehousing: Introduction to Data Mining and Data Warehousing. *Web Data Mining and the Development of Knowledge-Based Decision Support Systems*, 2017, pp. 312–337. https://doi.org/10.4018/978-1-5225-1877-8.ch016.
7. Garg, N., R. M. Flynn, F. Pecht, and D. R. Brill. FAA's Instrumentation Project at John F Kennedy International Airport to Study Load and Environment Induced Responses in Concrete Pavements. *Proc., 9th International Conference on the Bearing Capacity of Roads, Railways and Airfields*, Norway, 2013.
8. NOAA. *National Centers for Environmental Information*. https://www.ncdc.noaa.gov. Accessed April, 2017.
9. Al-Qadi, I. L., H. Wang, P. Yoo, and S. Dessouky. Dynamic Analysis and In Situ Validation of Perpetual Pavement Response to Vehicular Loading. *Transportation Research Record: Journal of the Transportation Research Board*, 2008. 2087: 29–39.
10. Al-Qadi, I. L., and H. Wang. Full-Depth Pavement Responses under Various Tire Configurations: Accelerated Pavement Testing and Finite Element Modeling. *Journal of the Association of Asphalt Paving Technologists*, Vol. 78, 2009, pp. 721–760.
11. Al-Qadi, I. L., A. Loulizi, M. Elseifi, and S. Lahouar. The Virginia Smart Road: The Impact of Pavement Instrumentation on Understanding Pavement Performance. *The Journal of Association of Asphalt Paving Technologists*, Vol. 83, 2004, pp. 427–466.
12. Al-Qadi, I. L, and S. Bhutta. In Situ Measurements of Secondary Road Flexible Pavement Response to Vehicular

Loading. *Transportation Research Record: Journal of the Transportation Research Board*, 1999. 1652: 206-216.
13. Al-Qadi, I. L, S. Portas, M. Coni, and S. Lahouar. Runway Instrumentation and Response Measurements. *Transportation Research Record: Journal of the Transportation Research Board*, 2010. 2153: 162–169.
14. Dong, M., G. F. Hayhoe, and Y. W. Fang. Runway Instrumentation at Denver International Airport: Dynamic Sensor Data Processing. *Presented at Airfield Pavement Conference*, Seattle, WA, 1997, pp. 363–378.
15. Orfanidis, S. J. *Introduction to Signal Processing*. Englewood Cliffs, NJ: Prentice-Hall, 1996.
16. MATLAB 2016a Documentation, "findpeaks", https://www.mathworks.com/help/signal/ref/findpeaks.html. Accessed April, 2017
17. MATLAB 2016b Documentation, "Prominence", https://www.mathworks.com/help/signal/ug/prominence.html. Accessed June, 2017
18. Stekhoven, D. J., and P. Bühlmann. MissForest—Non-Parametric Missing Value Imputation for Mixed-Type Data. *Bioinformatics*, Vol. 28, No. 1, 2011, pp. 112–118.
19. Larson, G., and B. Dempsey. *Enhanced Integrated Climatic Model Version 2.0*. Department of Civil and Environmental Engineering, University of Illinois at Urbana-Champaign, 1997.
20. Cortes, C., and V. Vapnik. Support-Vector Networks. *Machine Learning*, Vol. 20, No. 3, 1995, pp. 273–297.
21. Gunn, S. R. Support Vector Machines for Classification and Regression. *ISIS Technical Report*, No. 14, 1998, pp. 85–86.
22. Jebara, T. Multi-Task Feature and Kernel Selection for SVMs. *Proc., 21st International Conference on Machine Learning*, 2004, p. 55.
23. Al-Qadi, I. L., and H. Wang. *Evaluation of Pavement Damage due to New Tire Designs*. Illinois Center for Transportation (ICT), Rantoul, IL, 2009.
24. Timm, D. H., and A. L. Priest. *Wheel Wander at the NCAT Test Track*. Report No. 05-02, National Center for Asphalt Technology, Auburn, AL, 2005.

The ACRP Selection Panel for the Graduate Research Award Program on Public-Sector Aviation Issues peer-reviewed this paper (18-03064).

Article

TRR
JOURNAL OF THE TRANSPORTATION RESEARCH BOARD

Transportation Research Record
2018, Vol. 2672(29) 35–47

DOI: 10.1177/0361198118799710
journals.sagepub.com/home/trr
SAGE

A Risk Assessment Framework for Airport Development Projects

Anna Hopper[1]

Abstract

This paper develops a risk assessment framework for airport development projects. It discusses the major types of inherent *development risk*, including political risk, environmental risk, financial risk, airline risk, forecast risk, and regulatory or operational risk, and it offers suggestions for risk mitigation strategies. Furthermore, it identifies and analyzes *relative risk determinants*, which affect the magnitude and type of risk that development projects will likely face. These include the presence of a dominant airline, the airport's rate structure, the airport's ownership and operating structure, local demand, and geopolitical events. These factors and their interconnected relationships are illustrated through case studies of relevant airport development projects.

U.S. airports have many development needs. A recent Airports Council International report claims that in the next four years alone, U.S. airports will collectively require over US$100 billion in expenditures on capital improvements (*1*). Although these programs have the potential to expand capacity, improve services, and increase efficiency, history shows that they are likely to be over budget and behind schedule. Expectations are frequently wrong. ACRP Report 116 (*2*) cites research by Flyvbjerg, Bruzelius, and Rothengatter which found in a "study of 167 transportation infrastructure projects constructed over the previous 70 years [...] that large capital projects were initially underestimated by an average of 20%, with a small percentage of projects even being overestimated" (*2*, p. 3). These miscalculations are not surprising: "[s]uch projects are inherently risky because of long planning horizons and complex interfaces" (*4*, p. 579).

Before beginning a major development project, airport management and planners will want to distribute resources to minimize risk as much as possible. However, "risk allocation is wholly dependent on the adequacy of risk identification" (*5*, p. 4), a topic which has been underanalyzed in the area of airport development projects. This paper aims to fill that gap by focusing on the questions: what are the inherent risks in airport expansion projects, and how can airports better anticipate these risks?

The paper develops a risk assessment framework for development projects. This framework provides an outline of different types of *development risks* that projects may face and a method for evaluating the risks' relative importance through the discussion of *relative risk determinants*. It is a primer on understanding the risk landscape that may be useful not only to airport management and policy boards but also to other stakeholders involved in the process, particularly those new to the airport industry. It should be seen as a precursor to developing a project-specific risk management strategy.

Ten case study airports provide the foundation for the framework's development. These airports have experienced a range of developments projects, from runway additions to terminal projects to airport relocations. Though they illustrate major programs, similar risks may also occur with less comprehensive projects. Table 1 previews the airports and the specific development projects included in the analysis. All of the airports are located in the United States; within the country, they are geographically diverse. The cases are composed of medium to large airports because the effects of development are more easily observed; however, the lessons should be applicable to smaller airports as well

Based on these cases, the paper develops a typology of the major *development risk* categories that airport projects are likely to encounter. Risks are separated into two

[1]Harvard University, Cambridge, MA

Corresponding Author:
Address correspondence to Anna Hopper: ahopper@fas.harvard.edu; ahopper@post.harvard.edu

Table 1. Summary of Case Study Airports

Airport	Project type
DEN: Denver Int'l Airport	Airport (1990s)
MCI: Kansas City Int'l Airport	Airport (1970s), Terminal (ongoing)[a]
MSP: Minneapolis St. Paul Int'l Airport	Runway (1990s–2000s)
PIT: Pittsburgh Int'l Airport	Terminal (1990s–2000s)
ORD: Chicago O'Hare Int'l Airport	Runways (2000s–2010s)
RSW: Southern Florida Int'l Airport	Airport (1980s), Terminal (1990s–2000s)
SAN: San Diego Int'l Airport	Airport (2000s)[a]
SEA: Seattle-Tacoma Int'l Airport	Runway (1990s–2000s)
SFO: San Francisco Int'l Airport	Runway (1990s–2000s)[a]
STL: St. Louis Lambert Int'l Airport	Runway (1990s–2000s)

Note: Int'l = International.
[a]Project incomplete or abandoned.

Table 2. Types of Development Risk

Pre-completion risks	Post-completion risks
Political risk	Airline risk
Environmental risk	Regulatory/operational risk
Financial risk	Forecast risk
Airline risk	

Table 3. Relative Risk Determinants

Airport-specific	External
Presence of dominant airline	Local demand
Rate structure	Geopolitical events
Government structure	

primary categories: pre-completion and post-completion of the project. The specific types of risks discussed are previewed in Table 2 and are discussed in detail in the following sections.

Furthermore, the paper identifies factors termed *relative risk determinants*, or variables that affect the relative importance of different risk types. These risk determinants are also separated into two types: endogenous, or airport-specific, characteristics and exogenous, or external, factors. The different types of relative risk determinants are illustrated in Table 3 and described in further detail later in the paper.

The rest of the paper proceeds as follows. First is a brief descriptive overview of the case studies examined in this study. Then, each type of risk is discussed individually. Examples of the risks are drawn from the aforementioned case studies. The different relative risk determinants are subsequently explained, with a focus on how they relate to the various risk categories. Conclusions are presented at the end.

Case Study Preview

This section provides a brief introduction to the airports and development projects first referenced in Table 1. Subsequent sections will elaborate on these examples in greater detail to illustrate the different kinds of risks and relative risk determinants identified in this paper.

- DEN: Denver International Airport: In the 1970s, limited capacity and exhausted development potential at Stapleton Airport precipitated a decision to build an entirely new airport for Denver. In 1995, after legal battles, delays due to design changes, and cost overruns of over US$2 billion, Denver International Airport became the largest U.S. airport by land area (*6–8*).
- MCI: Kansas City International Airport: When it was originally built, the airport was designed to maximize traveler convenience by minimizing the distance between the curb and the plane. However, with the introduction of airport security screenings, the layout became extremely inefficient. The airport is still working through possible solutions to convert the airport to a single terminal design (*9–14*).
- MSP: Minneapolis St. Paul International Airport: The US$700 million project to build Runway 17/35 was the centerpiece of a US$3.2 billion expansion project (*15*). Planning started when the Metropolitan Airports Commission, directed by the Minnesota State Legislature, began considering options in 1989. Approved by the FAA in 1998, the runway opened in 2005 (*15–17*).
- ORD: Chicago O'Hare International Airport: Chicago's airspace is some of the busiest airspace in the country. To alleviate congestion at ORD, in 2005 the airport undertook an US$8 billion Modernization Program that included adding, removing, and lengthening runways to make six parallel runways and two diagonal runways (*18*). Although the project was extremely controversial, it has been quite successful at achieving its goals (*16, 18–22*).
- PIT: Pittsburgh International Airport: In the early 1990s, the Allegheny County Airport Authority spent approximately US$1 billion on the construction of a new midfield terminal to allow US Airways to expand its hub operations at the airport. It later ended up closing parts of the new terminal when US Airways' hub shut down (*23–26*).

- RSW: Southwest Florida International Airport: This airport in Fort Myers, FL was originally opened in 1983 as an alternative to the area's existing airport, Page Field. However, it outgrew its capacity far ahead of schedule, leading to the addition of a new midfield terminal that opened in 2005 (*27–31*).
- SAN: San Diego International Airport: Lindbergh Field is located in the middle of San Diego, which limits the potential for its expansion. As the traffic through SAN began to approach the airport's capacity, the San Diego Airport Authority considered overhauling the military airport at nearby Miramar and moving the airport there, though the move ultimately did not occur (*32–35*).
- SEA: Seattle-Tacoma International Airport: In the late 1980s, studies predicted that SEA could reach capacity by 2000. A large public involvement program concluded that although a new airport would not be possible, a third runway should be built. In 1992, plans began for the new runway, but opposition and environmental issues slowed them down. It did not open until November 2008, at which point the cost had risen from US$217 million to over US$1 billion (*36, 37*).
- SFO: San Francisco International Airport: Although SFO has two pairs of parallel runways, they are too close together for both to operate during bad weather. With no land around the airport suitable for expansion, additional runways would have to extend into San Francisco Bay. In the late 1990s and early 2000s, expansion proposals suggested filling in the Bay to create land for runways, but the plans were ultimately suspended due largely to environmental concerns (*38–43*).
- STL: St. Louis Lambert International Airport: This airport built a third parallel runway between 1998 and 2006. Though the capacity expansion was originally intended to support TWA's hub at STL, the runway has not been used as much as projected since TWA left (*15, 25, 44–48*).

Types of Risk

Pre-Completion Risks

Pre-completion risks are those that are most likely to occur during the planning or construction phases of a development project. They can be contrasted with post-completion risks, which are discussed next.

Political Risk. Political risk is the uncertainty associated with the potential actions of neighbors, constituents, or politicians who have a stake in the development process. It can arise when public or government approval is necessary for the project to proceed. As U.S. airports are publicly owned, city, state, or federal approval, or some combination of these, may be required at different stages of development. Political risk can also become an issue when anti-expansion movements or protests cause extreme delays and cost overruns. Since activists frequently cite environmental issues as a reason for their opposition, political risk may be closely related to environmental risk (see the subsequent section for details).

Examples of Political Risk

- In some cases, the future of the airport may be in the hands of constituent voters, as occurred in San Diego. At its 2003 creation, the San Diego Airport Authority was specifically tasked with studying potential new airport sites because there was limited potential for expansion at downtown Lindbergh Field (*32, 33*). After considering 32 potential locations, the authority proposed a controversial move to the Marine Corps Air Base at Miramar (*33*). They held a public referendum on the future of the airport (*35*). Many politicians and military brass spoke out against the plan. Residents near Miramar worried about the excessive noise that might arise from a shared civilian and military airfield. Business leaders were openly divided on the plan. Ultimately, the referendum failed, with 62% of the voters against the plans and only 37% in favor (*33*). Though the vote was technically nonbinding, its failure combined with the substantial outcry against Miramar caused the Airport Authority to abandon plans to move away from Lindbergh Field (*34*).
- A similar situation is ongoing in Kansas City with plans to convert MCI into a single terminal airport. The existing terminal design allows passengers to drive up right in front of their gates, which is convenient for passengers but causes problems for security screening and baggage control (*9*). The current renovation plans will achieve this goal by overhauling the existing Terminal A and closing the other two terminals. Due to a 2014 city law, Kansas City's City Council is required to hold a public vote to issue new airport revenue bonds or begin any significant airport construction projects (*11*). Unlike in San Diego, the business world is more strongly on the side of change, but opposition remains significant. Although polls showed that the city's voters were staunchly against a single terminal airport (*13*), a vote in late 2017 approved the renovation (*12*). The exact plans are still under development.

- Political risk can also come from organized opposition protest groups, as occurred at ORD in response to the airport's Modernization Program. In order to build new runways, the government utilized eminent domain to claim over 500 homes and businesses in nearby Elk Grove and Bensenville (*20*, *21*). The use of a former cemetery was particularly controversial, as 1,494 bodies had to be moved before construction could begin (*21*). Residents banded together to protest and mount a legal challenge against the takeover. Ultimately, settlements were reached out of court, but it was a protracted battle (*18–20*). Because eminent domain is frequently a tool for airports to gain the land necessary for expansion programs, it is particularly important to understand this type of political risk.

Takeaway: Airport development projects will likely face one or more of three main types of political risk, each of which is illustrated by the above examples. The first type of risk occurs when constituents have a direct referendum on a project. This was present both in San Diego and in Kansas City, though with differing outcomes. The second type is the risk that politicians will campaign or vote against a project as a way to curry favor among their constituents. San Diego also provides an example of this kind of activity. The third type of risk occurs when organized groups mount protests or legal challenges. As in the case of ORD, these have the potential to sideline the project for years.

Although it is not possible to avoid political risk, there are some strategies that can help mitigate it. The World Economic Forum's 2015 report on "Mitigation of Political & Regulatory Risk in Infrastructure Projects" suggests that both "inclusive community engagement" and "effective and constructive interaction with the public sector" should be priorities (*49*, p. 6). The cultivation of an open channel of communication and dialogue with a variety of stakeholders may help avoid worst-case scenarios. This communication should be a consistent priority, not only during the development process but also throughout the entire life of the airport.

Environmental Risk. Environmental risks are uncertainties related to environmental factors, regulations, and permits. Typical major issues include noise pollution, land conservation, and wildlife preservation. Laws often mandate environmental impact studies for major projects, and those studies may uncover unanticipated issues that make permitting more difficult. Moreover, environmental risk may be closely associated with political risk. For example, environmental activists or anti-expansion neighbors may use environmental concerns as a basis for challenging a development.

Examples of Environmental Risk

- Environmental variables can cause extensive delays. Seattle's third runway project is one of many examples where this has occurred. Critics of the original plan objected on the grounds that it would result in increased noise pollution and the destruction of the surrounding wetlands. Despite an "impressive array" of noise abatement plans and a proposal to restore or generate five acres of wetlands for each acre impacted, the Puget Sound Regional Council deemed the initial plans insufficient (*36*). These issues also provided the grounds for opposition groups to hire an environmental lawyer to fight against the project. Rounds of environmental evaluations and litigation continued for years. In the end, the runway was allowed, but the State Department of Ecology outlined stringent environmental requirements, including the creation of over 120 acres of wetlands (*36*).
- Environmental risk may also contribute to a project's demise, as occurred at SFO. Due to land constraints, San Francisco proposed filling in San Francisco Bay to create space for parallel runways extending into the water. If successful, this project would have involved dredging as much as 2.5 square miles of the Bay (*42*). However, environmental opposition was widespread. Numerous studies were mandated, and the outcry over the increased noise and potential impact on wildlife was intense (*43*). Following the 9/11 decrease in air traffic, the project was ultimately deemed unnecessary and halted (*41*). The airport is now again facing a serious capacity constraint (*38*).
- It can, however, be possible to manage environmental risk and potentially even find positive sides to it. RSW, for example, turned it into a tool for success. During its recent midfield terminal expansion project, the airport voluntarily proposed the conservation of a 7,000-acre swamp that abuts airport property (*31*). Though this "Mitigation Park" cost US$30 million to build and US$500,000 per year to maintain (*30*), its existence helped the airport avoid additional rounds of environmental studies and qualified it to apply for particular FAA grants, accelerating the approvals process and tempering the costs.

Takeaway: Nearly every major airport development project will have to deal with environmental risk, and planners should anticipate that mitigating environmental concerns may be very expensive. A common tactic for opponents involves drawing out the process until it eventually dies. Given the high probability of extreme delays, airports should attempt to alleviate concerns in advance as much as possible. RSW provides a good example of an airport that was able to do so successfully through the use of a voluntary conservation program.

Airline Risk. Airline risk is the risk related to airline approval or disapproval of the proposed plan. As major users of the airport's facilities, airlines are significant stakeholders in the development process. They will likely have strong opinions about potential development projects, and they may also wield substantial bargaining power. This is particularly true if they are expected to contribute to the project financially. The specific airport's rate structure may also affect the airline's position in the planning process because it can influence the cost- and revenue-sharing agreement between the airline and the airport. More on rate structure is given in the relevant section below.

Examples of Airline Risk

- Airlines can be major drivers of development projects. Trans World Airlines (TWA) originally urged Kansas City to adopt its existing decentralized design (*14*), but now many airlines advocate its conversion to a single terminal. Southwest in particular supports this idea because it would like to expand connecting traffic at MCI, as it has in nearby St. Louis (*11*). However, MCI's current terminal design is not conducive to that plan. Indeed, Southwest so strongly favored the single terminal idea that it not only pledged financial support but also vowed not to support any alternative plans that involve keeping and renovating the existing terminals (*10*, *11*). Though having airline support for the project can be beneficial, this hardline stance adds another dimension of stress to the planning process. MCI is not the only airport to deal with this issue. Airlines often have strong opinions, both in favor of and opposed to development plans, and Southwest is frequently vocal.
- Each airline is an individual stakeholder, so airlines may also have competing views on development opportunities, and those views must be managed. For example, in 2005 when SEA was trying to determine a solution to its capacity needs, both Alaska Airlines and Southwest balked at the potential for fee increases to fund a third runway (*36*). Instead, they each proposed building their own terminals and moving services to nearby Boeing Field. However, at that time the facility could not reasonably support both airlines given the required road improvements and noise abatement measures for such a dramatic increase in flights. The plans were ultimately vetoed entirely; neither airline received preferential treatment (*37*). The third runway was added to the existing SEA airport instead, but the situation resulted in the Port "comb[ing] through its budget [to] whittle down projected airline fees at Sea-Tac" (*37*).
- Airlines may also demand certain features as a condition of their support for a project, as United did when Denver built its new international airport. In June 1991, United signed a lease to be a dominant airline at the new airport but only after the airport agreed to install a rapid baggage handling system, a major development for airports at the time. United also changed the specifics of its requests several times during the process (*8*). For example, in August 1992, while the airline eliminated plans for the system to transfer bags, it added automated oversize luggage handling and a special ski equipment facility (*6*). Moreover, the system itself was plagued with problems that delayed the opening of the airport multiple times. The delays were very costly for both the airline and the airport, and the system was ultimately abandoned (*6*, *7*).

Takeaway: Airlines are key stakeholders in airport development projects, so there is inherent risk associated with their role in the process. They can be either project drivers or obstructionists. When disagreements between airlines and airports occur, they often center around three aspects of the project: 1) whether it should happen, 2) what it should look like, and 3) how much money it should cost the airline. To the extent possible, airports should choose partners carefully, though often there may not be multiple options.

Financial Risk. Financial risk concerns uncertainties about the overall cost of a development project. It comes from several sources, including uncertainties about how the project will be funded, delays leading to increased costs over time, and unknown construction variables. It may be related to other types of risk, such as regulatory risk, environmental risk, or airline risk.

Examples of Financial Risk

- SEA's third runway provides an example of how development costs can multiply over time. When the runway was first proposed in the 1992 Master Plan, its estimated cost was US$217 million (*36*). As planning proceeded, many environmental issues arose. By 1995, efforts to solve those problems had driven the cost up to US$430 million

(*36*). By its 2008 opening, the runway's total cost had reached over US$1 billion (*6*). The cost overruns were so high that the state investigated the Port of Seattle for wasteful spending.
- The new airport in Denver also faced cost overruns of over US$4 billion beyond initial estimates (*8*). However, in contrast with Seattle, the project was not delayed due to opposition. Rather, many of the expenses were due to the major changes in scope demanded by United Airlines, which were discussed in relation to airline risk. The scope changes not only increased the bill, but they also delayed the construction timetable, which had a cascading effect on other parts of the project (*6*, *8*).

Takeaway: Like all development projects, airport developments face different forms of financial risk. Resources for mitigating financial risk are numerous and parallel those in other industries (e.g., insurance), so they are not addressed in detail here. However, when assessing financial risk, airport planners should be aware that transportation projects are particularly vulnerable to "optimism bias" (*50*). This type of bias results in forecasts providing unrealistically low cost estimates and underestimating the risk of cost overruns. As financial analyses or forecasts for public infrastructure like airports may be "instruments for getting public funds committed to a favored project" (*4*, p. 587) [describing results from Wachs (*51*)], they should be regarded with a high level of scrutiny. If possible, quantitative methods ought to be employed to correct potential optimism bias. Examples of such corrections can be found in Flyvbjerg and COWI (*50*).

Post-Completion Risks

Because projects are often intended to serve the airport for decades after completion, the uncertainties associated with a development project do not end once the project is completed. This section describes several types of risk that may be relevant months or even years after the facility opens. In particular, topics of discussion include risk associated with airline stakeholders, risk of incorrect forecasts, and risk of unexpected regulatory or operational changes.

Airline risk. Post-completion airline risk is the uncertainty that an airline will change its service in ways that impact the airport's operations and profitability. This is often closely related to forecast risk (see the relevant section below). The airport is particularly vulnerable to airline risk if it relies heavily on a dominant airline that operates over half of the airport's flights or if an airline operates a hub at the airport. Pittsburgh and St. Louis are two of the most notable examples of dominant airlines de-hubbing after major development projects.

Examples of Post-Completion Airline Risk

- Between 1987 and 1992, Pittsburgh spent US$1 billion to build a new midfield terminal to increase capacity for its hub airline, US Airways (*23*). With its innovations in design, including an X-shaped terminal and the addition of an "AirMall," the project was lauded as a model modern airport. Of the 76 new gates, US Airways leased nearly three-quarters of them (*52*). The changes were supposed to allow US Airways to expand their operations at PIT. The airline agreed to bear a substantial amount of the cost by repaying US$608.7 million in bonds issued by Allegheny County (*52*). To cover this obligation, its airport fees, which had previously been some of the country's lowest, were increased substantially. However, when the events of September 11, 2001 precipitated a major downswing in air travel around the country, the US Airways hub at Pittsburgh was especially hard-hit. The high airport fees became even more difficult for US Airways to pay, and the airline began losing US$40 million per year on its PIT operations (*24*). In 2004, citing these financial concerns, the airline opted to close its hub at PIT. The airport quickly went from serving over 500 daily flights to serving fewer than 250 daily (*26*). Much of the once-new midfield terminal was sealed off or abandoned.
- St. Louis faced a similar scenario. As a major TWA hub since deregulation and the airline's headquarters since 1993 (*15*), STL was once the 10th busiest airport in the United States (*48*). During bad weather conditions, only one of the two existing close parallel runways could operate, resulting in major backlogs and delays (*46*). Moreover, as STL was a hub, those delays were felt across the country. In 1998, more than 20 years of airfield capacity limitation discussions finally resulted in approval of the US$1.059 billion Airport Expansion Program, which involved building a third parallel runway to serve the demand (*45*). However, TWA had been in and out of bankruptcy twice, in 1992 and 1995 (*44*). Despite a brief turnaround in the late 1990s, TWA again filed for Chapter 11, one day after it was sold to American Airlines in April 2001 (*45*). This was unfortunate timing as construction on the new STL runway began in July (*45*), and the devastating impacts of 9/11 on the airline industry quickly followed (*45*). American subsequently decided to drop STL as a hub to focus on its existing hub cities.

Takeaway: Airline risk is particularly worrisome when a development project is intended to serve the expansion needs of a single airline. If that airline is the airport's dominant airline or hub operator, the risk may be even higher. In these cases, development plans' cost-benefit analyses should include contingencies that assess the potential outcomes of the project if the dominant airline were to leave the airport. Furthermore, in some cases it may be worth considering strategies to diversify revenue streams in order to minimize exposure to a single airline's decisions. This could involve actively campaigning for additional carriers or routes from non-dominant carriers (*53*)—perhaps by offering incentives such as lower fees. However, it might also take the form of non-aviation revenue. Airports interested in this strategy should consult *ACRP Synthesis 19: Airport Revenue Diversification* (*54*) for more information.

Forecast Risk. Future passenger demand is very difficult to predict accurately. De Neufville and Odoni even argue that "the forecast is 'always wrong'" (*55*, p. 88). Forecast risk is the risk that expected demand estimates are so far off in direction, magnitude, or both that the airport can no longer operate as planned. This risk is particularly relevant for development projects because forecasts are frequently used both to justify the need for enhanced capacity and to determine the extent of development necessary. If forecasts are too optimistic, the airport risks spending millions on facilities that may not be used. Meanwhile, overly conservative forecasts may cause airports to under-build, resulting in over-crowding until new facilities can be added.

Examples of Forecast Risk

- Both PIT and STL forecasts far overestimated future demand. In Pittsburgh, the 1993 Master Plan anticipated serving 40.8 million travelers by 2013. However, largely due to the closing of US Airways' hub, the airport served only 8.3 million passengers in 2012. The same year, half of the A and B concourses in the midfield terminal were closed because only 50 gates (down from 104) were needed for regular operations (*23*). St. Louis faced a similar scenario as a TWA hub. Prior to the construction of the new runway, supplements to the airport's 1996 Master Plan projected that by 2015 the annual number of passengers served would be somewhere between 26 and 56 million with a baseline forecast of 42 million (*28*). Although the lowest estimate acknowledged that the airport might lose its position as a TWA hub, it assumed that another airline's hub would replace it. However, this replacement did not occur. In 2015 the airport served only 12.7 million passengers (*48*); it has not come close to former projections. Moreover, the runway built to enhance TWA's operations has not been utilized as much as anticipated; in 2009 it accounted for less than 10% of the airport's annual runway operations (*25*).
- Forecasts may also underestimate demand. Fort Myers, FL faced this scenario with RSW Airport in the 1980s when demand far exceeded planned capacity. When the airport opened in 1983, forecasts suggested it would reach its full capacity of 3 million passengers per year in 1995 (*27*). However, as Figure 1 shows, the airport actually reached that limit by 1988. The airport's design allowed only minor gate expansions, which was not enough to serve the increased demand. Though planning began for a new midfield terminal, it did not open until 2005, at which point the airport was carrying 8 million passengers per year, more than double the original facility's intended capacity (*30*, *31*).

Takeaway: Airport planners should begin a project with the understanding that the demand forecast will likely be incorrect. Unexpected changes, whether due to airline decisions or macroeconomic factors, may result in actual demand that is higher or lower than the forecast's margin of error. To mitigate this risk, airport planners will want to forecast a wide range of possible outcomes. As Kincaid, Tretheway, Gros, and Lewis argue, "forecasting must take into consideration what can happen in addition to what seems most likely to happen" (*53*, p. 7). Contingency plans should be developed for each of the potential scenarios. To the extent possible, the plans should emphasize flexibility in design. For example, a more adaptive terminal design at RSW that allowed for a sizable addition might have eliminated the immediate need for a new terminal building. In addition, Kincaid et al. (*53*) suggest identifying specific "trigger points"—moments in time when an event occurs or a threshold is reached that precipitates the next stage of development. For example, a trigger point might suggest that when the airport reaches a pre-determined number of annual passengers, work begins on another terminal.

Regulatory/Operational Risk. Airports are subject to a comprehensive collection of rules and standards created by outside organizations, including but not limited to the FAA, TSA, and Congress. The World Economic Forum's 2015 report on "Mitigation of Political & Regulatory Risk in Infrastructure Projects" explains that these regulations are often highly specific, and small changes can have significant impacts (*49*).

In the context of development projects, regulatory risk is related to changes in requirements, standards, or rules imposed by outside agencies or organizations that change

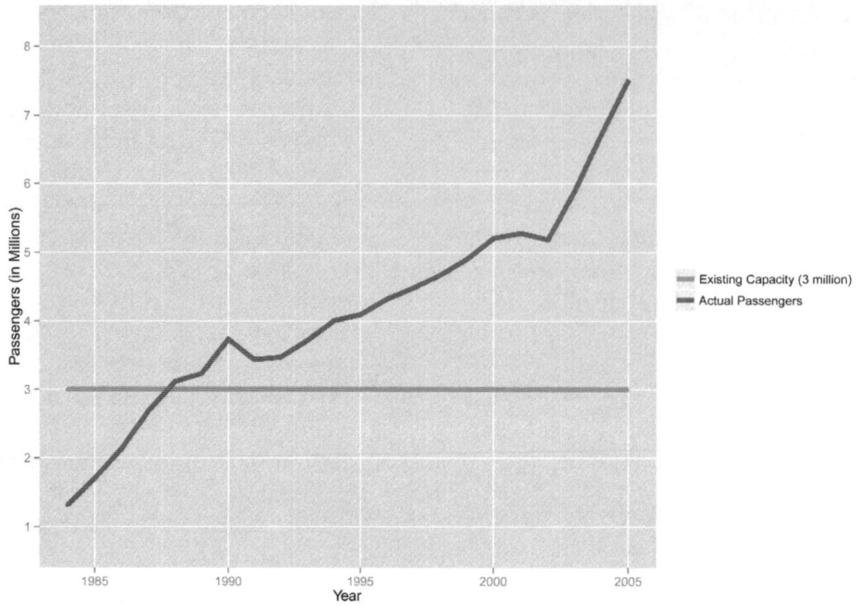

Figure 1. Growth vs. capacity at RSW (*30, 31*).

the operational effectiveness of the expanded facility after it has been built.

Examples of Regulatory/Operational Risk

- The increase in required airport security screening is one of the most significant regulatory changes that has affected airports in the last two decades. This issue has particularly challenged MCI because the airport's layout is not conducive to integrating screening facilities for all passengers. The gates are laid out along one side of curved terminals with the passenger pick-up/drop-off area on the other so that passengers can drive up within 75 ft of their gate. Architect Clarence Kivett explained to the *Kansas City Star* that it was created to be the "world's shortest walk-to-fly" airport (*9*). However, soon after the airport's 1972 opening, the rising number of airplane hijackings caused the federal government to mandate the introduction of passenger screening. Because there was no single entry point to gates, separate stations had to be located at nearly every gate. Increased security requirements post-9/11 magnified the problems (*9*). Not only did the new passenger screening equipment require additional space, but checked baggage screening was also mandated. Though most airports operate using a single, centralized luggage screening system, MCI's design and structure required the implementation of six separate systems (*14*). Though lauded as a futuristic airport at the time of its opening, MCI has not been able to effectively adapt that design to work with the changing regulatory requirements.
- MSP faced a different kind of regulatory and operational challenge with the addition of Runway 17/35. When it was built, the nearly US$700 million runway had two main purposes: 1) limit noise by sending 40% of loud departures across the Minnesota River and away from local residents and 2) raise the airport's capacity to 90 flights per hour (*15*). Two years following the runway's opening, the actual number of departures via Runway 35 was nearly 40% lower than planned (*15*). In part, this result can be attributed to lower-than-anticipated demand levels (related to forecast risk), but FAA decisions also limited the operational success of both major objectives. The runway's location required aircraft to cross over another runway, so when there was a choice, air traffic control frequently opted not to utilize it. These flow decisions were reinforced by airlines themselves who had a vested interest in avoiding the longer taxi times necessitated by the runway crossing. As a result, the noise reduction for surrounding residents was much lower than intended. Moreover, on August 5, 2015, the FAA suspended all arrivals on Runway 35 (*15*). Despite the fact the agency had originally approved the runway's location, it later decided that the potential for

collisions with departures on nearby Runway 30L was too high. This change effectively reduced the airport's maximum capacity per hour by about 30% from the original projection (*16*). Due to these post-construction limitations, along with less demand than expected, MSP's new runway did not have its intended impact (*15*, *16*).

Takeaway: Airport regulations happen in many forms. MCI and MSP provide examples of major regulatory changes that impacted airport operations. However, the World Economic Forum (*49*) points out that "[a]ny small change to the [regulatory] details—to permissible noise levels, for example, [...] can have a hugely detrimental effect on revenues or cost" of an infrastructure project (p. 15). As with forecast risk, this type of risk highlights the need for planners to anticipate possible change and to emphasize flexibility and adaptability of design.

Relative Risk Determinants

Given the typology of development risks presented above, airports must evaluate which ones are most relevant or important for the project under consideration. The following section identifies relative risk determinants, variables that influence the incidence of one or more types of risks. They are divided into both endogenous factors, which are related to characteristics of the airport, and exogenous factors, which are outside variables that influence the airport. A solid understanding of the relationships between the development risks and relative risk determinants will help airports determine which risks are most likely to affect their project. A mapping of relationships between risk types and relative risk determinants is shown in Figure 2.

Airport-Specific Risk Determinants

Airport-specific risk determinants are directly related to characteristics of the airport itself. They cannot easily be changed, so airport management and planning teams will normally have to work within their confines to determine the most feasible solutions.

Presence of a Dominant Airline. The presence of a dominant airline at an airport can significantly increase airline risk, forecast risk, and financial risk. Airports in this situation include both large hub airports and smaller airports with a single carrier that operates more than 50% of flights. When an airport relies heavily on one airline, it may expand to support that airline's continued service. However, such a unilateral focus leaves the airport vulnerable if the airline suddenly cuts flights, whether due to changes in network strategy or to unexpected financial circumstances. This can, in turn, result in dramatic departures from previous demand forecasts and financial predictions. Kraus and Koch point out that "[e]specially during crisis periods—which are experienced quite frequently in the volatile aviation industry—airlines have to react [...with] significant cost-cutting measures at short notice. Airports can suffer substantially" (*56*, p. 99). They may be left with expensive and unused infrastructure.

In the best-case scenario, the local demand is strong enough that other airlines will replace the one that left. However, for hub airports in particular, this may not be the case. Burghouwt explains that "[h]ubs allow metropolitan areas to grow beyond the size of their local market" (*57*, p. 9). As a result, if the dominant airline closes its hub, the remaining local demand may be insufficient to support the airport's traffic levels. As previously discussed, this situation occurred in both Pittsburgh and St. Louis. Though these were both large hub airports, smaller airports with a dominant airline should not overlook this possibility.

Rate Structure. Rate structure can be an important relative risk determinant because it may influence the amount of voice that airlines have in the planning process. Rate structures can be either compensatory, residual, or hybrid (*1*, *58*). In a compensatory structure, airports assume the primary risk of running the airport but also receive the most revenue if it goes well; airlines pay only their share of costs. In a residual structure, airlines pay not only their own costs but also a fraction of the remaining residual costs after other airlines have also covered their own costs; however, they also have the opportunity to share in the residual profits, if any exist. Hybrid structures combine features of compensatory and residual structures. As airlines have a greater stake in the financial health of the airport under a residual cost scheme, they may have more of a say in major airport development projects, thus influencing the airport's exposure to both airline risk and financial risk.

Government Structure. The government structure and rules may also affect the relative importance of certain risks. Because U.S. airports are government-owned, the ownership structure may dictate the decision-making process and stakeholder involvement. As a result, it can certainly influence political risk because it may affect which voices can be heard. Environmental regulations and required steps during development projects may also be dictated by the government structure. It can additionally influence financial decision-making and therefore financial risk. Finally, the structure may determine the process by which financial decisions about development projects are made. For example, municipally run airports may be

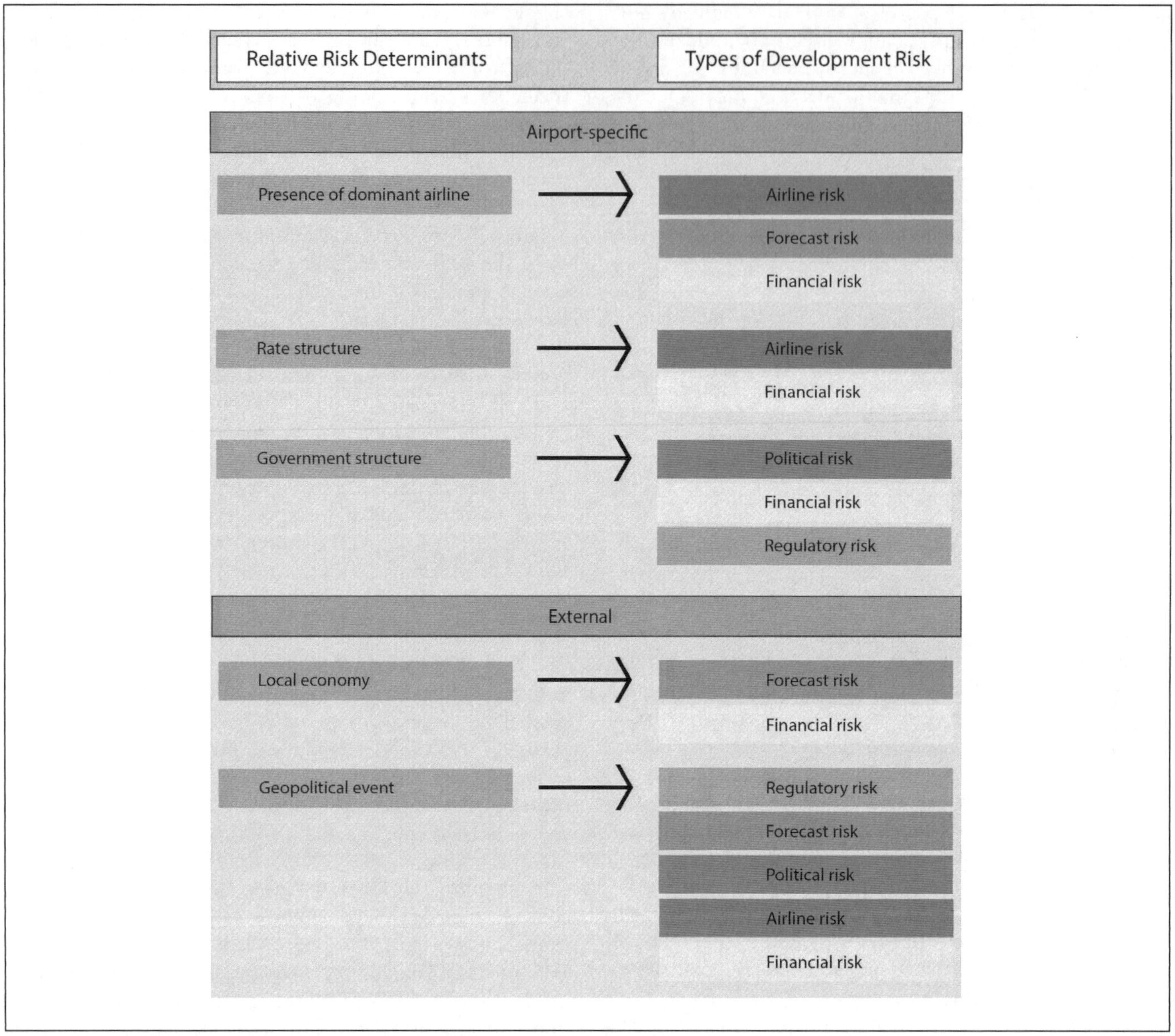

Figure 2. Relationships between relative risk determinants and risk types.

more directly responsible to constituents for financial decisions than airports run by independent agencies.

External Risk Determinants

These kinds of risk determinants are factors not specific to the airport and are usually outside of the airport's control, but they may affect the relative importance of different development risks.

Local Demand. An airport located in an area with very strong local demand may be less susceptible to forecast and airline risk. Two factors are important. First, with strong local demand, airlines are less likely to suddenly cut or eliminate service to the airport. Second, even if a major airline does leave, the level of demand is more likely to incentivize other airlines to fill the empty slots. However, the converse is true if local demand is weak. Then, the airport is much more vulnerable to service changes. It is more critical that the fate of the airport does not hinge on a single airline, as occurred at STL and PIT.

Geopolitical Events. Of all forms of transportation, air travel is particularly affected by major geopolitical events. Though they are often unanticipated, these do have the potential to affect nearly every type of risk to a project, including airline risk, forecast risk, financial risk, political risk, and regulatory or operational risk. These kinds of events include, for instance, fuel crises, wars, or terrorist attacks. For example, RSW was in the middle stages of its new terminal renovation when the 9/11 attacks prompted a need for increased security measures and explosives screening for baggage (*29*). The airport had to revise its design plans mid-construction in order to meet these new regulations.

Conclusion

Taken together, the development risks and relative risk determinants presented in this framework highlight important relationships that influence airport improvement projects. The paper begins by identifying and describing seven major types of risk inherent in airport developments, including political risk, environmental risk, financial risk, pre-completion airline risk, forecast risk, post-completion airline risk, and regulatory/operational risk. Case studies provide real-life examples of airports that have dealt with these different risks, and some suggestions for risk mitigation are offered.

Following this discussion, the paper introduces five kinds of relative risk determinants and their relationship with the various development risks. The relative risk determinants include the presence of a dominant airline, the airport's rate structure, the ownership and operating structure of the airport, local demand, and geopolitical events. These factors affect the magnitude and kind of risks that a particular development project is likely to face. The potential links between the different kinds of relative risk determinants and risk types are illustrated in Figure 2.

Airport managers, planners, policy boards, and other stakeholders should be able to use the information in this paper to better assess what risks are most relevant for their own airport. Understanding the landscape of risks should allow them to develop a project-specific risk management strategy as a next step. *ACRP Report 116: Guidebook for Successfully Assessing and Managing Risks for Airport Capital and Maintenance Projects* would be a useful reference (*2*). Quantitative airport-specific risk assessments are also recommended. *ACRP Report 76: Addressing Uncertainty about Future Airport Activity Levels in Airport Decision Making* provides an introduction to some potential forecasting strategies, including Monte Carlo simulation (*53*).

Aside from this paper's discussion of specific development risks and their relative importance, readers should come away with an understanding of the benefit of open dialogue with both supporters and opponents, as well as the importance of flexibility in planning and the need to broaden the scope of anticipated scenarios in order to more accurately predict airport development outcomes.

Acknowledgments

Thanks to the ACRP Graduate Research Award Program, especially Larry Goldstein of the Transportation Research Board, Dave Byers of Quadrex Aviation, Kent Duffy of the FAA, and Kitty Friedheim of Friedheim Consulting for invaluable guidance and extremely helpful feedback throughout the process of writing this paper. Thanks also to Richard de Neufville of MIT, Tony Gomez-Ibanez of the Harvard Kennedy School, and Daniel Ziblatt of Harvard University for their support.

References

1. Airports Council International - North America. *Airport Infrastructure Needs 2017-2021*. Online Fact Sheet. aci-na.org/sites/default/files/2017airportinfrastructureneedsonepager.pdf.
2. Price, M. *ACRP Report 116: Guidebook for Successfully Assessing and Managing Risks for Airport Capital and Maintenance Projects*. Airport Cooperative Research Program, Transportation Research Board of the National Academies, Washington, D.C., 2014.
3. Flyvbjerg, B., N. Bruzelius, and W. Rothengatter. *Megaprojects and Risk: An Anatomy of Ambition*. Cambridge University Press, 2003.
4. Flyvbjerg, B. Policy and Planning for Large Infrastructure Projects: Problems, Causes, Cures. *Environment and Planning B: Planning and Design*, Vol. 34, 2007, pp. 578–597.
5. Nielsen, K. R. Some Practical Thoughts—Risk Allocation Regarding Airport Projects in China. *Proc., IPBA Conference*, Risk Allocations on Airports Session, Beijing, China. 2007.
6. Calleam Consulting, Ltd. *Case Study - Denver International Airport Baggage Handling System - An Illustration of Ineffectual Decision Making*. 2008. https://www5.in.-tum.de/~huckle/DIABaggage.pdf.
7. de Neufville, R. The Baggage System at Denver: Prospects and Lessons. *Journal of Air Transport Management*, Vol. 1, No. 4, 1994, pp. 229–236.
8. Dempsey, P. S., A. R. Goetz, and J. S. Szyliowicz. *Denver International Airport: Lessons Learned*. McGraw-Hill, New York, 1997.
9. Hendricks, M. The Why of KCI: A Broken Plan that Many Travelers Still Love. *The Kansas City Star*. March 8, 2014. http://www.kansascity.com/news/politics-government/article341740/The-why-of-KCI-A-broken-plan-that-many-travelers-still-love.html.
10. Horsley, L. Airlines to Kansas City: Get on Board for New KCI Single Terminal. *The Kansas City Star*, April 26, 2016. http://www.kansascity.com/news/politics-government/article73988477.html.
11. Horsley, L. Regarding KCI's Future, City Hall Ponders a New Flight Path. *The Kansas City Star*, June 24, 2016. http://www.kansascity.com/news/local/article85755692.html.
12. Rodriguez, L. Kansas City Voters Overwhelmingly Approve New Single Terminal at KCI. *KCUR 89.3*, November 7, 2017. http://kcur.org/post/kansas-city-voters-overwhelmingly-approve-new-single-terminal-kci#stream/0.

13. Rose, S. What If a New KCI Never Gets off the Ground? *The Kansas City Star*, December 17, 2016. http://www.kansascity.com/opinion/opn-columns-blogs/steve-rose/article117497103.html.
14. Yeager, M. Hidden Problems Cause Headaches for Airlines at KCI. *KSHB Kansas City*, August 14, 2012. http://www.kshb.com/news/local-news/investigations/hidden-problems-causing-headaches-for-airlines-at-kci.
15. Terrell, K. MSP Airport Planning: Let's Not Rush Toward a Future We'll Hate. *Minneapolis Star Tribune*, August 15, 2015. http://www.startribune.com/msp-airport-planning-let-s-not-rush-toward-a-future-we-ll-hate/322555041/.
16. Doyle, P. In Wake of MSP Runway Suspension, Lawmakers Demand Further Analysis of Airport Expansion Plans. *MinnPost*. September 30, 2015. https://www.minnpost.-com/politics-policy/2015/09/wake-msp-runway-suspension-lawmakers-demand-further-analysis-airport-expansi/
17. *Master Plan Supplement Study: Lambert-St. Louis International Airport*. Final Report. Leigh Fisher Associates, 1996.
18. Ruthhart, B. Chicago, Airlines Nearing $8.5 Billion Deal to Dramatically Expand O'Hare. *Chicago Tribune*. February 26, 2018. https://www.chicagotribune.com/news/local/politics/ct-met-city-hall-story-20180223-story.html
19. *Community Uprooted: Eminent Domain. In the Way of Progress: Bensenville and the O'Hare Airport Expansion*. Loyola University Chicago. http://www.luc.edu/eminent-domain/siteessays/bensenvilleil/. Accessed December 17, 2018.
20. Dodge, J. As O'Hare Expanded, Time Lapse Shows Demise of Bensenville Neighborhood. *2 CBS Chicago*. July 25, 2014.
21. Placek, C. Dispute Over Cemetery Land for Runway Comes to an End. *Daily Herald*, December 28, 2012. http://www.dailyherald.com/article/20121228/news/712289857/.
22. Schaper, D. O'Hare to Get 6th Runway, But Without Expanded Terminals, Delays May Continue. *NPR*, February 1, 2016. http://www.npr.org/sections/thetwo-way/2016/02/01/ 465101435/ohare-will-get-a-sixth-runway-but-without-expanded-terminals-delays-may-continue.
23. Belko, M. Pittsburgh International Airport's Midfield Terminal at 20: A Shell of Its Past Self. *Pittsburgh Post-Gazette*. http://www.post-gazette.com/news/transportation/2012/09/30/Pittsburgh-International-Airport-s-midfield-terminal-at-20-A-shell-of-its-past-self/stories/201209300226. Published September 30, 2012.
24. Grossman, D. Dismantling Pittsburgh: Death of an Airline Hub. *USA Today*, October 15, 2007. https://usatoday30.usatoday.com/travel/columnist/grossman/2007-10-15-dismantling-pittsburgh-hub_N.htm.
25. Leiser, K. Little-Used Runway Pros, Cons: Airport Says It's A Selling Point; Former Neighbors Still Angry. *St. Louis Post-Dispatch*. October 14, 2009. https://www.stltoday.com/news/little-used-runway-pros-cons-airport-says-it-s-a/article_364a3e80-6365-568e-9222-c9a9963642cd.html
26. Sharkey, J. Pittsburgh, Once a Showplace Hub, Feels US Airways Woes. *New York Times*, September 14, 2004. http://www.nytimes.com/2004/09/14/business/pittsburgh-once-a-showplace-hub-feels-us-airways-woes.html.
27. Board, P. T., and E. B. Colcord. *A History of Aviation in Lee County, Florida*. http://www.flylcpa.com/uploads/pagesfiles/511.pdf. Published May 14, 1993.
28. Lee County Port Authority Department of Communications and Marketing. *Southwest Florida International Airport (RSW) Total Passengers*. http://www.flylcpa.com/uploads/pagesfiles /2379.pdf.
29. Mayk, L. Fort Myers Airport Work on Course. *Herald Tribune*, September 12, 2002. http://www.heraldtribune.com/news/20020912/fort-myers-airport-work-on-course.
30. Southwest Florida International Airport. *FAST FACTS*. http://www.flylcpa.com/uploads/pagesfiles/2230.pdf. Accessed November 29, 2018.
31. Southwest Florida International Airport. *Mitigation Park*. http://www.flylcpa.com/mitigation/. Accessed December 17, 2018.
32. Davis, R. Airport Questions Answered. *The Voice of San Diego*, April 20, 2006. http://www.voiceofsandiego.org/uncategorized/airport-questions-answered/.
33. Davis, R. Miramar: What Went Wrong? *The Voice of San Diego*. November 9, 2006. http://www.voiceofsandiego.org/2006/11/09/miramar-what-went-wrong/.
34. Halverstadt, L. For San Diego Businesses, the Sky is What's Limiting. *The Voice of San Diego*, September 14, 2014. http://www.voiceofsandiego.org/business/for-san-diego-businesses-the-sky-is-whats-limiting/.
35. Proposition A: Commercial Airport at MCAS Miramar, San Diego County Regional Airport Authority. *SmartVoter.org*, November 7, 2006. http://www.smartvoter.org/2006/11/07/ca/sd/prop/A/.
36. Crowley, W. *Sea-Tac International Airport: Third Runway Project*. HistoryLink.org, 2003. http://www.historylink.org/File/4211. Updated January 30, 2010.
37. Langston, J., and G. Holt. Plan Won't Fly: Sims Kills Southwest's Boeing Field Hopes. *Seattle Post-Intelligencer*, October 11, 2005. http://www.seattlepi.com/local/article/Plan-won-t-fly-Sims-kills-Southwest-s-Boeing-1184887.php.
38. McGinnis, C. SFO Prepares for Slots as More Flights Crowd In. *SF Gate*. May 30, 2018. https://www.sfgate.com/chris-mcginnis/article/SFO-slot-contols-FAA-12948559.php
39. San Francisco International Airport. *Five-Year Capital Plan: Fiscal Year 2015/16 - Fiscal Year 2019/20*. http://media.flysfo.com/media/sfo/about-sfo/capital-plan-fy1516_1.pdf. Accessed July 1, 2015.
40. San Francisco International Airport. *SFO Officials React to Independent Report on Alternatives to Reduce Delays without Runway Separation*. http://media.flysfo.com/CRArelease2_24_01.pdf.
41. SFO Airport Runway Expansion Put in Holding Pattern. *USA Today*, May 16, 2003. http://usatoday30.usatoday.com/travel/news/2003/05/16-sfo-expansion.htm.
42. Sward, S. Critics Question Impartiality of SFO Runway Study/Feinstein's Husband Part-Owner of Firm Conducting Report and has 6% Stake in Northwest Airlines. *San Francisco Chronicle*, June 12, 2000. http://www.sfgate.com/bayarea/article/Critics-Question-Impartiality-of-SFO-Runway-Study-2755460.php.

43. Wilson, M. Roar of Disapproval for SFO Runway Expansion Plans. *San Francisco Chronicle*, July 31, 2001. http://www.sfgate.com/bayarea/article/Roar-of-disapproval-for-SFO-runway-expansion-plans-2893602.php.
44. Grant, E. X. TWA - Death of a Legend. *St. Louis Magazine*. July 18, 2006. https://www.stlmag.com/TWA-Death-Of-A-Legend/
45. Lambert St. Louis International Airport. *The Expansion Story?* https://web.archive.org/web/20080501193743/http://www.lambert-stlouis.com/e/newwebsite/id261.asp. Archived webpage from May 1, 2008.
46. Mutzabaugh, B. New $1 Billion Runway Opens this Week, But It's Not Needed Anymore. *USA Today*, April 11, 2006. https://web.archive.org/web/20070830212520/http://blogs.usatoday.com/sky/2006/04/st_louis.html.
47. Stoller, G.. St. Louis' Airports Aren't Too Loud: They're Too Soft. *USA Today*, January 9, 2007. https://usatoday30.usatoday.com/money/biztravel/2007-01-09-st-louis-usat_x.htm.
48. Strong Passenger Traffic at Lambert Pushes More Growth in 3rd Quarter. St. Louis Lambert Media Release. October 27, 2016. https://www.flystl.com/newsroom/stl-news/2016/strong-passenger-traffic-at-lambert-pushes-more-growth-in-3rd-quarter
49. World Economic Forum and The Boston Consulting Group. *Mitigation of Political & Regulatory Risk in Infrastructure Projects: Introduction and Landscape of Risk*. 2015.
50. Flyvbjerg, B., and COWI. *Procedures for Dealing with Optimism Bias in Transport Planning: Guidance Document*. UK Department for Transport, London, 2004.
51. Wachs, M. Ethics and Advocacy in Forecasting for Public Policy. *Business and Professional Ethics Journal*, Vol. 9, 1990, pp. 141–157.
52. Marriott, M. Pittsburgh Builds Airport of the Future Now. *New York Times*. November 12, 1991. https://www.nytimes.com/1991/11/12/us/pittsburgh-builds-airport-of-future-now.html
53. Kincaid, I., M. Tretheway, S. Gros, and D. Lewis. *ACRP Report 76: Addressing Uncertainty about Future Airport Activity Levels in Airport Decision Making*. Transportation Research Board of the National Academies, Washington, D.C., 2012.
54. Kramer, L. S. *ACRP Synthesis 19: Airport Revenue Diversification - A Synthesis of Airport Practice*. Transportation Research Board of the National Academies, Washington, D.C., 2010.
55. de Neufville, R., and A. Odoni. *Airport Systems: Planning, Design, and Management* (2nd ed.). New York, McGraw Hill, 2014.
56. Kraus, A., and B. Koch. The Opportunities and Threats of Turning Airports into Hubs. *Journal of Air Transportation*, Vol. 11, No. 1, 2006, pp. 87–112.
57. Burghouwt, G. Airport Capacity Expansion Strategies in the Era of Airline Multi-hub Networks. International Transport Forum Discussion Papers No. 2013/05. OECD Publishing, Paris, 2013. http://dx.doi.org/10.1787/5k46n421b15b-en.
58. Alfred, R. Jr., and D. E. Starcher. *Contract Risk Management for Airport Agreements*. Airport Cooperative Research Program Legal Research Digest 30. The National Academies Press, Washington, D.C., 2016.

The ACRP Selection Panel for the Graduate Research Award Program on Public-Sector Aviation Issues peer-reviewed this paper (18-01544).

TRR
JOURNAL OF THE TRANSPORTATION RESEARCH BOARD

Article

Transportation Research Record
2018, Vol. 2672(29) 48–58

DOI: 10.1177/0361198118799708
journals.sagepub.com/home/trr
SAGE

Air Cargo Forecasting in an Age of Electronic Retail

Peter Hylton[1] and Catherine L. Ross[1]

Abstract

Airports face numerous difficulties in capital planning as a result of instability and unpredictable demand. Some trends such as excess bellyhold capacity have depressed demand for dedicated cargo flights, whereas others like growth in electronic retail (e-retail) are stimulating demand unevenly among airports. If e-retail distribution is following a different pattern from general cargo, then it may reshuffle cargo volumes among airports in a way that is unforeseeable by most forecasting approaches. It has been difficult to estimate whether e-retail shipments would follow the same geospatial patterns as general cargo because of the dearth of data on e-retail distribution networks. This study examines domestically oriented e-retail logistics to test its relationship with airport cargo throughput controlling for regional variables that also influence cargo volume. The regression models show that an airport's accessibility to landside e-retail logistics activity helps predict that airport's cargo volume in a way that access to traditional retail logistics does not. The findings support the alternative hypothesis that e-retail may reconfigure distribution networks and boost cargo volume at some airports, opening opportunities to gain market share for airports that are new to e-retail. Depending on the extent of the reconfiguration, e-retail growth could temporarily degrade the accuracy of common forecasting methods. Airports can respond by collecting qualitative and quantitative data where possible and adopting scenario-based, adaptive policy making that hedges risks.

Even though the air cargo industry has seen weakening domestic demand over the past decades and integrator consolidation around a few hub airports, cargo remains a driver of many airports' activity and budgets (*1*). Bellyhold cargo generates revenue for passenger airlines that helps make their passenger segments profitable, particularly on Pacific and other international routes where yields are higher and large wide-body aircraft have greater capacity (*2*). At some airports, yields and volumes justify domestic integrator flights or international freighter flights that produce aeronautical revenue and bolster surrounding warehouse lease rates (*3*). In the right conditions, air cargo can support airport budgets and provide connections that serve regional industries. Air cargo may also stimulate regional or national economic growth (*4*, *5*), and vice versa (*6*, *7*). Capital investments' long timeline puts a great responsibility on airport planners to accurately predict the timing and approximate scale of passenger and cargo demand. Even in ideal conditions, this is difficult, and the task becomes even more muddied when aviation undergoes qualitative transitions.

Electronic retail (e-retail) is one of the forces affecting aviation. It generates freight and mail demand at international gateways (*8*, *9*), stimulates business for integrators (*10*), and even spurred the creation of Amazon Prime Air, a cargo airline dedicated to Amazon, the world's largest e-retailer. Airport Cooperative Research Program (ACRP) Report 143 characterized electronic commerce, which includes e-retail, as one of the four large forces shaping air cargo (*1*). Nonetheless, very little data has been available to quantify e-retail logistics and to characterize its configuration, and the trend's recent ascendancy has impeded efforts to understand the nature and impacts of e-retail's demand for air cargo service.

This study is one of the first attempts to empirically examine associations between regional e-retail logistics and airport cargo volumes, and to consider changes to forecasting methods to improve their accuracy in an age of e-retail. It examines creating a new independent variable for e-retail logistics facility locations that could be used as an alternative to e-retail air cargo data, as logistics facilities in an airport region can potentially furnish air cargo shipments and stimulate demand. It compares

[1]City and Regional Planning, Georgia Institute of Technology, Atlanta, GA

Corresponding Author:
Address correspondence to Peter Hylton: peter.hylton@gatech.edu

e-retail logistics with brick-and-mortar (B&M) logistics to test the hypothesis of a statistically significant correlation between e-retail logistics and airport cargo that is not present with B&M logistics. The study also proposes guidance for airports to enhance their ability to accommodate e-retail.

The paper includes six sections including the introduction. The literature review explains how e-retail is related to aviation and catalogs airports' forecasting approaches. The methodology section outlines the analytical methods used in the study and describes the major data inputs. The results section presents an assessment of whether e-retail is associated with airport cargo in a way that may not be explained by more commonly used factors. The discussion section contains an evaluation of four air cargo forecasting approaches and their utility in anticipating e-retail cargo, modifications that might improve accuracy, and data needs. Lastly, the conclusions section draws overall lessons for airport management and suggests future research directions.

Literature Review

The literature review addresses three primary topics. There is an examination of e-retail's transportation needs to better understand how e-retail moves goods and why logistics facilities sometimes locate near airports. The review includes a description of retail logistics operations in order to identify factors motivating outbound transportation facility location choice. Lastly, air cargo forecasting approaches are categorized to probe their sensitivity to new dynamics like e-retail that are absent from historical data.

E-retail logistics assumes many forms. In some cases, buyers and sellers are in different countries, triggering shipments by land, air, or sea depending on the value of the items being shipped and the urgency of the delivery. International e-retail may involve fulfillment centers in the United States serving overseas customers, or imports from overseas retailers serving the American market. By some estimates, international e-retail shipments could account for 20% of global air cargo volumes (*11*), with integrators and freight forwarders providing transportation (*9*, *11*, *12*). International e-retail logistics appears to be oriented around gateway seaports and airports, often using ground transportation to meet domestic transportation legs (*13*). For example, an American e-retailer might move goods purchased online by foreign customers by truck several hundred miles from a fulfillment center to an international gateway airport. Then the goods could be loaded aboard a freighter or wide-body passenger aircraft for transport to the customers' country of residence. From there, ground or air transport might be used depending on costs and delivery urgency. Whether domestic or international, air shipment is more expensive than surface modes, so air is normally only used when demand was not predicted far enough ahead to use surface modes exclusively.

Theory suggests that e-retail logistics facilities will locate differently than B&M facilities, and that airport access may be one of the factors that influence e-retail logistics facilities. Domestically oriented e-retailers' distribution networks must accommodate swift delivery of small, short-notice, and unpredictable shipments to many addresses. Some fulfillment centers are located near the large consumer markets that they serve, whereas others are farther from large markets but central to integrator networks (*14*). E-retailers use a mix of transportation modes for fulfillment based on cost and delivery requirements. E-retail's compressed delivery time frame and geospatially dispersed demand (i.e., centered on customers instead of stores) makes demand harder to predict and therefore results in more situations in which surface transportation is unable to meet delivery schedules. Accordingly, e-retail's growth is expected to increase overnight cargo flights (*15*).

The empirical literature has examined trends in transportation and freight facility location, although it has almost never distinguished e-retail from B&M retail. Airports have been observed to attract different types of logistics activity, both within and among metropolitan regions, as evidenced by higher warehouse rents (*14*, *16*), manager preferences (*17*), and facility proximity (*18*, *19*). Logistics facilities sometimes locate near airports to postpone drop-off time with carriers for overnight delivery (*20*). Highways (*16*, *21*), uncongested roads (*22*), blue-collar labor pools, low-cost commuting options, low-regulation environments (*17*), and in some cases seaports and railroads (*16*, *17*, *23*) also appear to attract logistics facilities. The authors are not aware of any quantitative work connecting airports' cargo volumes and nearby e-retail logistics that provides evidence about how e-retail could alter specific airports' activities.

The existing body of knowledge suggests that the e-retail sales channel's growth will galvanize demand at some airports, particularly at airports that are very central to air cargo transport networks. Airports that are especially attractive to e-retail cargo may wish to consider the sector in their plans and forecasting models. Airport plans need to forecast future cargo volumes, and they can use at least three methods: time series trend analysis, regression-based econometric modeling, and market share analysis. For brevity, they will not be detailed here, but ACRP Report 143 provides specifics on their application (*1*). In practice, many airport plans supplement one of these three approaches with multiple scenarios (*24*). Scenario planning allows forecasts to consider changes in key factors that would bring very different futures.

This study employs a cross-sectional econometric model. An econometric model should include as many potentially explanatory variables as possible from several of these categories (*25*). Macroeconomic variables (e.g., gross domestic product, composite leading indicators, and jet fuel prices) (*26*, *27*), socio-demographics (e.g., population, income), the airport's infrastructure-constrained capacity, and carrier networks and fleet plans can all shape cargo volumes (*1*). Some models use regional economic variables (e.g., total personal income) as predictors (*28*), and several studies have found regional labor markets to be related to air cargo volume (*29*), especially employment in retail, wholesale, and manufacturing (*30*). Similar research has associated economic development with airport cargo, airport capacity, highway access, other nearby transportation infrastructure, and surrounding labor markets (*29*, *31*). Factors such as airport cost, operational flexibility, and infrastructure can shape cargo volume indirectly through carrier hub choice and capacity (*32*).

E-retail is likely to be one of the major forces shaping air cargo growth, and the literature also suggests that e-retail has an especially large influence on airport activity at some cargo hubs (*8*, *33*). Nonetheless, e-retail has not had a major place in airport planning, partially because the sector is still new and also because e-retail shipment data is hard to obtain. To date the literature has not contained studies quantitatively examining whether airport cargo is associated with accessibility to e-retail logistics facilities in a way that it is not for B&M logistics, nor described adaptations of airport forecasting approaches to control for an e-retail effect if one is present. This study seeks to contribute to the body of knowledge and aspires to fill gaps in the literature.

Methodology

This study explores the relationship between accessibility to e-retail logistics facilities and airports' cargo volume. The study proposes that airport access to e-retail logistics facilities associates with air cargo volume and that B&M access will not have a comparable relationship. It further proposes that airports may benefit from new data types or adapted forecasting methods to model e-retail's influence on their activity. The search for associations recognizes bidirectional influences among airports, shippers, and carriers (*4*, *34*). Cargo refers to express freight, non-express freight, and mail. The analysis includes five major steps: retailer and airport selection, data collection, variable calculation, model execution, and model interpretation. Based on the presence or absence of associations, potential adaptations to airport planning and forecasting methods are presented.

There have been two primary challenges hindering studies of the relationships between e-retail and airports. The first is the speed at which e-retail logistics is changing, making studies of its configuration capable of capturing only snapshots of a rapidly evolving industry. This limitation remains, although it can be controlled by couching quantitative studies in distribution strategy. The second challenge is the dearth of data to analyze e-retail shipments, which this study addresses by leveraging logistics facility locations as a placeholder for flows and imputing potential connections with airport activity based on proximity and access. It is one of the first studies to attempt to broadly catalog U.S. e-retail logistics facilities and develop a measure of e-retail access that quantifies the e-retail logistics industry's locational accessibility to a given airport. These two innovations underlay this analysis.

Step 1: Select Airports

This study examines the 127 airports in the contiguous United States with over 3 million pounds annual transported freight weight according to the Bureau of Transportation Statistics' T-100 database (*35*). The threshold was established to produce enough airport regions for statistical analysis and eliminate airports that have little or no commercial cargo.

Step 2: Gather Data

Data were collected from government agencies (e.g., U.S. Census Bureau, Bureau of Transportation Statistics), private datasets (e.g., ReferenceUSA, Airports Council International, Woods & Poole), and airports themselves as described in Table 1.

Retailers are selected to represent either the e-retail sales channel or the B&M sales channel based on the proportion of their revenue generated by either. The only firms included in the sample owe more than 80% of their sales value to just one sales channel (*38*). All of their logistics facilities are attributed to their dominant sales channel except when a professional publication (e.g., local newspaper or trade publication) describes specific facilities' functions, thereby allowing facility-level attribution. Any facilities that do not hold inventory (e.g., sortation centers) are excluded. The final sample includes 31 B&M retailers with 257 distribution centers totaling nearly 19 million square meters and 29 e-retailers with 168 fulfillment centers representing almost 12 million square meters of capacity. The largest B&M retailers by facility area are Walmart, Target, Lowe's and JC Penney, and the largest e-retailers by area are Amazon, Walmart eCommerce, Target, and The Home Depot.

All facilities are in the contiguous United States. Initial facility locations were established through ReferenceUSA's dataset. Gaps in ReferenceUSA's data

Table 1. Variables and Data Sources

Name	Description	Data source
Freight (dependent variable)	2015 annual freight processed by airport (metric tons)	Bureau of Transportation Statistics, T-100 2015
Mail (dependent variable)	2015 annual mail processed by airport (metric tons)	Bureau of Transportation Statistics, T-100 2015
Cargo (dependent variable)	2015 Freight and mail combined, including originating, destination, and transit shipments (metric tons)	Airports Council International – North America, 2015 Traffic Report and Airport Websites
E-retail access (test variable)	$E - retail\ access = \sum_{i \in I} \frac{area_i}{\sqrt{dist_i}}$ where *area* is the facility floor area in square meters; *dist* is the Euclidean distance in kilometers to the airport centroid; and *i* is a given fulfillment center from the whole set *I*.	Calculated
B&M access (test variable)	$B\&M\ access = \sum_{j \in J} \frac{area_j}{\sqrt{dist_j}}$ where *area* is the facility floor area in square meters; *dist* is the Euclidean distance in kilometers to the airport centroid; and *j* is a given distribution center from the whole set *J*.	Calculated
Gross regional product (control variable)	2015 gross regional product in same metropolitan statistical area as airport	Woods & Poole
Population (control variable)	2015 population in same metropolitan statistical area as airport	American Community Survey, 5-year (2011–2015)
Median household income (control variable)	2015 income in same metropolitan statistical area as airport	American Community Survey, 5-year (2011–2015)
Manufacturing employment (control variable)	2014 manufacturing employment in same metropolitan statistical area as airport	American Community Survey, 5-year (2011–2015)
Retail employment (control variable)	2014 retail employment in same metropolitan statistical area as airport	American Community Survey, 5-year (2011–2015)
Wholesale employment (control variable)	2014 wholesale employment in same metropolitan statistical area as airport	American Community Survey, 5-year (2011–2015)
Curfew (control variable)	Binary variable that is "1" if airport has a curfew	Boeing (*36*)
Highway density (control variable)	Interstate centerline length (km) within 25 km of airport	Calculated
Catchment area (control variable)	Population accessible by overnight ground, based on mean overnight delivery radius for UPS and FedEx (399-km radius)	American Community Survey, 5-year (2011–2015)
Industrial land value (control variable)	Average value per warehouse square meter within 10 km of airport ZIP code	FindTheData (*37*)

were supplemented with news articles, Google Maps, corporate websites and other data sources. Each facility location was verified with aerial imagery, and assigned companies were checked against signs visible in street-view images from Google Maps. The retailer sample accounts for distribution from American retailers to American customers or, in some cases, overseas customers. It does not include foreign retailers or overseas facilities.

Step 3: Calculate Variables

Variables were selected to mimic econometric models as described in the literature review or approximate other links between airports and logistics facilities. Most variables relate to regional economies, and others focus on ground accessibility and airport operations.

All data are valid for year 2015 or the closest available year. Three dependent variables are calculated, each with a slightly different focus. Freight is "property, other than express and passenger baggage transported by air" (*39*). The second dependent variable is mail volume, which refers to U.S. Postal Service (USPS) surface shipments carried by air on a space-available basis through mail contracts aboard many different airlines (*39*). The exclusions of express freight and airmail postage are potentially problematic because many e-retail parcels are transported as express shipments. Therefore, a third dependent variable is included to compensate for the deficit. The third dependent variable is airport cargo tonnage, which includes all freight and mail together, including express freight and airmail postage. It was obtained from the Airports Council International – North America for 43 airports, and from airport websites for the 60 airports from the sample that report it. Cargo data were obtained for 103 airports, which reduces statistical power slightly compared with the full sample. Together, these three dependent variables provide a comprehensive review of cargo moving on various air carriers under different classifications. The three dependent variables are approximations subject to inevitable data reporting differences among airports and airlines. The variables are especially sensitive to the largest carriers' reporting methods. Even though FedEx is the largest domestic mail and freight carrier (*40*), it appears to mix mail and freight in its reported statistics based on activity at its largest hub at Memphis International Airport (*41*). As the FedEx Corporation's mail contracts (all modes and services) made up 66% of the value of airline mail contracts in 2015 (*42*), there may be very substantial mail undercounting and equal freight over-counting. The cargo variable helps control for this divergence by including both mail and freight in a single number.

The variables that allow for hypothesis testing rely on gravity models that explain the airports' accessibility to e-retail and B&M logistics facilities. Gravity models create a single number that summarizes each airport's proximity to the entire set of facilities of a given sales channel. Facilities have more "weight" in the number when they are larger or closer to the airport. A gravity model was selected over a radius-based threshold because metropolitan land markets experience other economic pressures and geographic limitations that can relegate logistics functions to a greater distance from the airport than in a sparsely developed region. The research question references e-retail, but B&M retail is also included to allow comparison between the emerging and established sales channels. All variables are detailed in Table 1.

Step 4: Create Model to Predict Airport Cargo Volume

A total of six models were created for the different pairings of the three dependent variables with the two test variables. Models 1, 2, 4, and 5 explain the freight and mail dependent variables. For these models, the remaining independent variables were input into a backward stepwise regression whereby statistically insignificant variables are removed from the model until only statistically significant variables remain. The models allow for testing the alternative hypothesis that e-retail access is more associated than B&M access with air freight and mail.

Models 3 and 6 test air cargo on the reduced set of 103 airports with available cargo data. The sample provides less statistical power than the full dataset, so all test and explanatory variables were input without backward stepwise removal. Additionally, correlations were run between the two gravity models and airport cargo to naively quantify the relative magnitude of associations.

Step 5: Assess Hypotheses

Hypothesis assessment depends on results' alignment across the different models. The null hypothesis can be rejected if e-retail access is significantly associated with several types of air cargo while B&M logistics is not statistically significant or is less associated as evidenced by regression coefficients.

Results

The logistics facilities' locations were analyzed geographically to assess the extent of clustering before the models were run. The e-retail logistics facilities are more geospatially clustered and located closer to airports than the B&M logistics facilities. For example, 32% of B&M facility floor area is within 25 km of the 127 airports under study, whereas just over half (52%) of the e-retail facility floor area is within 25 km of these airports. Some of the airports with the highest e-retail access host integrator hubs. All of the largest cargo airports have greater normalized e-retail access than B&M access.

Ordinary least squares regression assumptions were tested for each model, and the models met them well. Histograms and q-q plots showed the residual errors of the natural log (LN) of the dependent variables to be approximately normally distributed. Homoscedasticity was tested and confirmed in each model by examining the configuration of residual error plots. Gross regional product (GRP), population, retail employment, and wholesale employment are highly correlated, so all except the first were excluded from the models to avoid multicollinearity. Descriptive statistics for all variables are shown in Table 2.

E-retail access is significantly associated with airport freight and mail (models 1 and 2 respectively, as shown in Table 3). The other retained variables describe regional economics and accessibility. A higher GRP correlates with increased freight and mail volumes, which is expected because GRP reflects the size of the regional economy that the airport serves. Manufacturing employment is associated with freight, but not mail, which is less likely to transport industrial items. The size of the catchment area population is negatively associated with both freight and mail volumes, which is unexpected. However, catchment area population could be inadvertently

Table 2. Descriptive Statistics

	Mean	Median	Minimum	Maximum	Standard deviation
Freight (metric tons)	97,477	21,318	1,439	2,189,132	257,334
Mail (metric tons)	3,539	201	0	40,097	7,671
Cargo (metric tons)	244,724	54,884	219	4,290,638	582,781
E-retail access (index)	375	361	244	709	82
B&M access (index)	595	595	401	918	98
Gross regional product (millions of U.S. dollars [USD])	124,790	53,342	3,106	1,137,543	211,903
Population (people)	2,677,182	1,135,734	41,892	19,979,950	4,089,830
Median household income (USD)	55,374	53,534	33,266	95,405	9,999
Manufacturing employment (jobs)	76,879	33,733	604	501,325	106,638
Retail employment (jobs)	129,054	59,002	1,568	935,456	189,198
Wholesale employment (jobs)	59,857	20,736	442	463,231	99,843
Curfew (binary)	0.09	0	0	1.00	0.29
Highway density (km)	115	106	0	292	61
Catchment area population (people)	25,714,455	24,592,290	1,345,696	63,242,402	15,502,411
Industrial land value (USD/ft^2)	24	30	0	86	21

Table 3. Electronic Retail Access Model Results

	Model 1: coef.	Model 1: SE	Model 2: coef.	Model 2: SE	Model 3: coef.	Model 3: SE
Dependent variable	LN_freight		LN_mail		LN_cargo	
R^2	0.46		0.46		0.48	
N	127		127		103	
Constant	0.65	1.05	−21.32***	2.86	−1.33	1.53
E-retail access	4.1E-03**	1.70E-03	0.01*	4.60E-03	4.7E-03**	2.2E-03
LN_gross regional product	0.39**	0.15	2.33***	0.24	0.54**	0.20
LN_manufacturing employment	0.41***	0.15	na	na	0.46*	0.23
Curfew	−0.89**	0.39	na	na	−1.07**	0.48
SQRT_highway density	na	na	na	na	0.01	0.06
Catchment area population	−3.1E-08***	9.40E-09	−8.7E-08***	2.50E-08	−3.1E-08**	1.2E-08
Industrial land value	0.01*	0.01	na	na	0.01*	0.01

Note: LN = natural log; SQRT = square root; SE = standard error; coef. = coefficient; na = not applicable.
*$p<.10$. **$p<.05$. ***$p<.01$.

standing in for an omitted third variable representing a diversion of supra-regional freight from air to ground modes. The presence of an airport curfew negatively associates with freight, whereas higher industrial land values are reflected in higher freight volumes, perhaps because of a likely positive correlation between land value and the regional industry's size. The models predicting freight and mail have fairly high explanatory value, capturing approximately half of the data's variance. The overall finding from models 1 and 2 is that e-retail activity in an airport region is associated with higher freight and mail volumes at that airport. E-retail activity is a viable factor to include in air cargo forecasts as it is robust to different specifications.

Models 4 and 5 (shown in Table 4) include B&M access as an independent variable, and the results contrast strongly with the e-retail models. The stepwise regression excluded B&M retail logistics from models 4 and 5, meaning that there is not a statistically significant association between B&M access and airport freight or mail. GRP and manufacturing employment are positively associated with freight, just as in Model 1, whereas the presence of an airport curfew and catchment area are negatively associated. Model 4 describing mail requires fewer variables, retaining only GRP and catchment area population.

Models with cargo as the dependent variable (models 3 and 6) show e-retail access to be more highly associated with air cargo volume than is B&M access. In Model 3, e-retail access's coefficient is positive and statistically significant at the 95% confidence level, whereas B&M access has a smaller coefficient and only becomes significant when standards are reduced to the 90% confidence level. Control variables mimic the sign and direction that they assumed in the other models. To help confirm the findings, two-tailed Pearson correlation coefficients were calculated relating e-retail access and B&M access with the natural log of airport cargo tonnage. E-retail access

Table 4. Brick-and-Mortar Access Model Results

	Model 4: coef.	Model 4: SE	Model 5: coef.	Model 5: SE	Model 6: coef.	Model 6: SE
Dependent variable	LN_freight		LN_mail		LN_cargo	
R^2	0.42		0.45		0.47	
N	127		127		103	
Constant	1.07	0.99	−18.91***	2.5	−1.97	1.80
B&M access	na	na	na	na	3.6E-03*	2.1E-03
LN_gross regional product	0.41***	0.15	2.32***	0.24	0.57***	0.21
LN_manufacturing employment	0.50***	0.15	na	na	0.45*	0.24
Curfew	−0.99**	0.39	na	na	−1.12**	0.48
SQRT_highway density	na	na	na	na	0.02	0.06
Catchment area population	−2.2E-08***	8.1E-09	−6.1E-08***	2.0E-08	−3.2E-08**	1.3E-08
Industrial land value	na	na	na	na	0.02**	0.01

Note: LN = natural log; SQRT = square root; SE = standard error; coef. = coefficent; na = not applicable.
$^*p<.10$. $^{**}p<.05$. $^{***}p<.01$.

is correlated with airport cargo tonnage at the 99% confidence level (coefficient: 0.280; significance: 0.004), whereas the relationship is much weaker in coefficient and significance for B&M access (coefficient: 0.172; significance: 0.082). This suggests a stronger association between e-retail access and air cargo than is present for B&M access.

Taken together, the results indicate that e-retail access likely has a connection with air cargo that B&M access does not fully replicate. This finding supports the literature in the belief that e-retail may shape cargo throughput at specific airports in addition to industry-wide (*1*). The models do not assert that e-retail access is causing air cargo growth at specific airports per se; airport factors such as cargo connectivity could also influence the nearby logistics industry, and the literature has shown bidirectional causation between regional characteristics and airport traffic. Nonetheless, there are moderately strong patterns that align with theory and support the alternative hypothesis.

Discussion

The nature of e-retail cargo raises three questions for airport planning: (A) can current methods forecast e-retail cargo as accurately as they do other kinds of cargo? (B) What, if any, methodological adaptations are required? And (C) what data are necessary? These three questions are summarized in Table 5. The three forecasting approaches are regression-based econometric, time series trend extrapolation, and market share analysis. Scenario planning is discussed concurrently because of its widespread use in airport plans even though it can accommodate many forecasting methods.

Conceptually, there is no reason that the approaches will necessarily forecast e-retail cargo either more or less accurately than general cargo over the long run. Nor are methods such as econometric models or trend extrapolation capable of predicting hub closures or other extra-model airline dynamics that confound forecasts (*24*, *43*). Nonetheless, there are at least three dynamics that will make it difficult for any of the approaches to accurately forecast e-retail cargo volumes in the near term, and one or more operate in all of the approaches. The first dynamic is that e-retailers have not yet established a financially sustainable, widely accepted distribution model. Some e-retailers are transitioning from nationally oriented distribution networks to regional and increasingly local distribution (*44*), and many are still too small to achieve substantial economies of scale. The industry's instability will bedevil forecasts in the short term. The second short-term confounding dynamic is e-retail's recent ascension, which provides a much abbreviated number of years from which to draw data for trend extrapolation. Trend extrapolation will have lower accuracy for e-retail than more established cargo types until a longer-term record can be constructed. The third challenge is the potential for mode shift due to e-retailers' preference for cheaper surface transportation. When e-retailers are relatively small and new, they are still learning the nature of consumer demand, which results in more occasions in which demand cannot be forecast adequately for ground delivery, requiring air transport instead. As they grow, e-retailers have more experience with consumer demand, which may allow them to reduce their reliance on air. Moreover, large e-retailers can build more facilities closer to customers, which may reduce the number of markets where air shipment is suitable. Therefore, uncertainties in e-retailers' future mode share will complicate airports' forecasts.

Even when these dynamics are resolved, each approach will likely benefit from some adaptations to forecast e-retail cargo as accurately as general cargo. The most readily adaptable approach is scenario planning.

Table 5. E-Retail in Forecasting Approaches

	Scenario	Regression-based econometric	Time series trend extrapolation	Market share analysis
A. Forecast accuracy	Lower until industry finds equilibrium	Lower until industry finds equilibrium and track record of data is built	Lower until industry finds equilibrium and track record of data is built	Lower until industry finds equilibrium
B. Adaptations	Build scenarios for high/low e-retail	This study suggests e-retail may not be entirely captured in existing models	See if e-retail cargo and other cargo are following the same trends	See if e-retail cargo and other cargo are following the same trends
C. Data requirements	Interviews with carriers and e-retailers	Present and past e-retail volumes at airport being forecast	Present and past e-retail volumes at airport being forecast	Present and future national e-retail volumes, and present e-retail volumes at airport being forecast

Many airports already create one or more scenarios that assume high-to-moderate growth (e.g., hub expansions, favorable economic conditions) and others that assume low growth or decline (e.g., hub closure, economic recession, high fuel costs). E-retail can simply be added to these scenarios, with the high-growth scenarios assuming regional development of an e-retail logistics hub and the decline scenarios assuming its absence. This study provides reasons to believe that regression-based econometric models may need new variables to adequately forecast e-retail cargo. The adaptations of trend extrapolation and market share analysis depend on whether e-retail cargo follows the same dynamics as other cargo types over time (i.e., trend extrapolation) or space (i.e., market share analysis). If the dynamics are different, then models could be split, with one model forecasting general cargo and a specialized model forecasting e-retail cargo.

Inadequate data is perhaps the greatest challenge to accurately forecasting e-retail cargo, and the data requirements become more difficult moving from left to right in Table 5. Scenarios can be created based on qualitative data that come from interviews with carriers or e-retailers about their shipments, future distribution networks, and the airport's place in those networks. The data requirements would be very similar to those that airports use today to generate forecasting scenarios. Regression-based econometric and time series trend extrapolation both require present and historical e-retail cargo data for the airport being forecast as a dependent variable. Individual airports may possess estimates for their e-retail cargo volume, but they are not widely available today. The most data-intensive approach is the market share analysis because it requires several local and national data as inputs: current e-retail cargo volume at the airport being forecast, current national e-retail cargo volume to estimate present market share, forecast national e-retail cargo volume, and a future market share if different from present. None of these data are available today, and individual airports will find national-level factors harder to estimate than local-airport factors. Therefore, approaches that use qualitative data (i.e., scenarios) or exclusively local information (i.e., econometric and trend extrapolation) have the most feasible data requirements.

In the short term, e-retail looks set to amplify uncertainty at airports. Airports can gather data and hedge to reduce their risks in an uncertain business environment. Data about e-retail and other logistics trends is useful whether quantitatively built into forecasting models or qualitatively considered in scenarios. For instance, airport staff can conduct interviews with local e-retailers, economic developers, or chambers of commerce to understand e-retail logistics' place in their region. They might also explore the possibility of obtaining estimates of e-retail cargo volumes from the cargo carriers serving the airport, such FedEx, UPS, and USPS, which may be able to identify e-retail shipments based on shipper name. Although data can decrease uncertainty, hedging acknowledges the remaining uncertainty and controls for risks. For instance, scenario planning tests an airport's response to best- and worst-case scenarios through high- and low-volume forecasts. Adaptive policy making is a formalized method for lowering risk by identifying steps that can be taken if selected risks become reality (*45*).

Airports that have little or no e-retail cargo may find new opportunities to attract it. One reason for the change is the e-retail industry's maturation, resulting in more logistics nodes and airports served. Another is the relative standardization of e-retail sales tax, which is putting states on a more level playing field for e-retail logistics. E-retailers are only required to pay sales tax in states where they have a tax nexus, which in state and federal law historically hinged on a physical presence. Conservative definitions of a tax nexus made states with low sales taxes or small customer bases especially attractive locations for e-retail logistics. Although still not 100% consistent, many states have passed legislation lowering the bar for establishing a tax nexus, making it much harder for e-retailers to avoid paying sales tax in any state (*46*, *47*). The result is a gradual breakdown in differences among e-retailers' tax obligations by state. For instance, in early 2017 Amazon began charging sales tax in all states, even in states where it has no facilities (*48*). Therefore, airports that are relatively near large customer bases may wish to consider means to cater to e-retail air cargo even if they currently serve little e-retail air cargo. Possible methods that merit further study include proactively meeting carrier requirements, the availability of landside or airside cargo facilities, attractiveness of aeronautical fees, and even partnerships with local colleges to develop needed workforce skills.

Conclusion

Airports are in the precarious situation of having to make capital investment commitments many years ahead of a demand that is uncertain. E-retail compounds uncertainty in an air cargo industry already undergoing rapid change. As e-retail captures more of the retail market share, air transport appears likely to serve some of its transportation needs for domestic and international shipments.

This study finds that e-retail logistics access is often associated with higher cargo volumes even controlling for regional and ground transportation factors, which suggests that e-retail may reconfigure air cargo networks and grow air cargo volumes. Moreover, the e-retail facilities' location indicates much greater clustering around airports, especially those with cargo hubs, than is seen for traditional retail logistics. As improved e-retail data become available, it may be possible to analyze cargo flows rather than facility locations as done here to understand specific airport's roles in e-retail distribution.

For many airports, e-retail is still too small a force to have a large effect on forecasts or operations, and for some airports this situation will remain for the foreseeable future. However, e-retail logistics is heavily concentrating in space and both theory and observation indicate that e-retailers turn to central airports for deliveries that cannot be made by other modes. Airports that are embedded in e-retail distribution networks are already experiencing formative influence on their cargo outlook, and even those that are now on the periphery of e-retail distribution networks may feel repercussions as the sales channel matures. These airports face additional uncertainty due to the dearth of data on e-retail cargo flows and the sector's fast-changing state, which accentuates their planning risk. To mitigate risk, they can gather qualitative data about their regional e-retail industry through business interviews and develop relationships with carriers that might provide visibility into the types of cargo using the airport. They can also build scenario-based and adaptive strategies to hedge risks. For now, the task for airports is to continue monitoring forces, including e-retail, that could deviate from previous forecasts and adjust projects accordingly while keeping an ear to the ground for opportunities that e-retail might bring.

Author Contributions

The authors confirm contribution to the paper as follows. Study conception and design: PH, CLR; data collection: PH; analysis and interpretation: PH, CLR; draft manuscript preparation: PH, CLR. All authors reviewed the results and approved the final version of the manuscript.

References

1. Maynard, M., D. Clawson, M. Cocanougher, D. Walter, R. Brimble, M. Webber, R. Janisse, K. Freidheim, and R. Miller. Airport Cooperative Research Program, Transportation Research Board, National Academies of Sciences, Engineering, and Medicine. *Guidebook for Air Cargo Facility Planning and Development*. ACRP Report, No. 143, 2015. https://doi.org/10.17226/21906.
2. WorldACD Market Data. *Yield Index Regions*. 2017. http://www.worldacd.com/index.
3. Ryan, R. Cargo: A Necessary Evil for Airports and Carriers? *Journal of Airport Management*, Vol. 2, No. 2, 2008, pp. 132–136.
4. Chang, Y.-H., and Y.-W. Chang. Air Cargo Expansion and Economic Growth: Finding the Empirical Link. *Journal of Air Transport Management*, Vol. 15, No. 5, 2009, pp. 264–265. https://doi.org/10.1016/j.jairtraman.2008.09.016.
5. Kasarda, J. D., and J. D. Green. Air Cargo as an Economic Development Engine: A Note on Opportunities and Constraints. *Journal of Air Transport Management*, Vol. 11, No. 6, 2005, pp. 459–462. https://doi.org/10.1016/j.jairtraman.2005.06.002.
6. Mosbah, S., and M. S. Ryerson. Can US Metropolitan Areas Use Large Commercial Airports as Tools to Bolster Regional Economic Growth? *Journal of Planning Literature*, Vol. 31, No. 3, 2016, pp. 317–333. https://doi.org/10.1177/0885412216653100.
7. Green, R. K. Airports and Economic Development. *Real Estate Economics*, Vol. 35, No. 1, 2007, pp. 91–112. https://doi.org/10.1111/j.1540-6229.2007.00183.x.
8. King, L. *New Schiphol Program a Boon for E-Commerce Shippers*. Air Cargo World, Seattle, WA, 2016.
9. *DHL Centre Taps into Asian E-Commerce Boom*. AirCargoNews. http://www.aircargonews.net/news/forwarders/single-view/news/dhl-centre-taps-into-asian-e-commerce-boom.html. Accessed March 13, 2017.

10. Kauffman, C. *E-Commerce, TNT Acquisition Boost FedEx Express in Q1*. Air Cargo World, Seattle, WA, 2016.
11. van Marle, G. Boom in Cross-Border E-Commerce Traffic Could Be a Lifeline for Air Cargo. *The Load Star*, London, UK, February 27, 2017.
12. Stevens, L., and E. E. Phillips. Amazon Plans Air Cargo Service for Chinese Customers. *The Wall Street Journal Online*, March 15, 2017.
13. Jiao, Z. Modes and Development Characteristics of China's Cross-Border E-Commerce Logistics. In: B. Liu, L. Wang, S. Lee, J. Liu, F. Qin, and Z. Jiao, eds.), Contemporary Logistics in China. Springer, Heidelberg, Germany, pp. 211–232. https://doi.org/10.1007/978-3-662-47721-2_12.
14. JLL. *United States PAGI Airport Outlook - 2015*. http://www.us.jll.com/united-states/en-us/research/5794/US-PAGI-Airport-2015-JLL. Accessed April 14, 2017.
15. Hesse, M. Shipping News: The Implications of Electronic Commerce for Logistics and Freight Transport. *Resources, Conservation and Recycling*, Vol. 36, No. 3, 2002, pp. 211–240. https://doi.org/10.1016/S0921-3449(02)00083-6.
16. Sivitanidou, R. Warehouse and Distribution Facilities and Community Attributes: An Empirical Study. *Environment and Planning A*, Vol. 28, No. 7, 1996, pp. 1261–1278. https://doi.org/10.1068/a281261.
17. Jakubicek, P., and C. Woudsma. Proximity, Land, Labor and Planning? Logistics Industry Perspectives on Facility Location. *Transportation Letters*, Vol. 3, No. 3, 2011, pp. 161–173. https://doi.org/10.3328/TL.2011.03.03.161-173.
18. Verboon, H. M., and E. Braun. *Clustering around International Airports*. Bachelor's thesis. Erasmus University, Rotterdam, the Netherlands, 2010.
19. Karsner, D. Aviation and Airports: The Impact on the Economic and Geographic Structure of American Cities, 1940s-1980s. *Journal of Urban History*, Vol. 23, No. 4, 1997, pp. 406–437.
20. Kasarda, J. D., and G. Lindsay. *Aerotropolis: The Way We'll Live Next*. Farrar, Straus and Giroux, New York, NY, 2011.
21. Cidell, J. Concentration and Decentralization: The New Geography of Freight Distribution in US Metropolitan Areas. *Journal of Transport Geography*, Vol. 18, No. 3, 2010, pp. 363–371. https://doi.org/10.1016/j.jtrangeo.2009.06.017.
22. Warffemius, P. M. J. *Modeling the Clustering of Distribution Centers around Amsterdam Airport Schiphol: Location Endowments, Economies of Agglomeration, Locked-in Logistics and Policy Implications*. Doctoral dissertation. Erasmus University, Rotterdam, the Netherlands, 2007
23. Bowen, J. T. Moving Places: The Geography of Warehousing in the US. *Journal of Transport Geography*, Vol. 16, No. 6, 2008, pp. 379–387. https://doi.org/10.1016/j.jtrangeo.2008.03.001.
24. Jacobs. *Executive Summary: Cincinnati/Northern Kentucky International Airport 2035 Master Plan Update*. 2013. https://www.cvgairport.com/docs/default-source/master-plan-report/2-exec-summary.pdf?sfvrsn=2. Accessed December 15, 2017.
25. Allen, P. G., and R. Fildes. Econometric Forecasting. In J. S. Armstrong (ed.), *Principles of Forecasting*. Springer, New York, NY, pp. 303–362. https://doi.org/10.1007/978-0-306-47630-3.
26. Wadud, Z. Simultaneous Modeling of Passenger and Cargo Demand at an Airport. *Transportation Research Record: Journal of the Transportation Research Board*, 2013. 2336: 63–74.
27. Mott MacDonald. *Appendix 2.1. Air Traffic Forecasting Methodology by IATA*. http://www.epd.gov.hk/eia/register/report/eiareport/eia_2232014/html/Appendix%202.1.pdf. Accessed April 2, 2017.
28. Jacobs Consultancy. *Master Plan Update: Portland International Airport*. City of Portland Bureau of Planning and Sustainability, Portland, OR, 2010.
29. Allroggen, F., and R. Malina. Causal Relationships between Airport Provision, Air Traffic and Economic Growth: An Econometric Analysis. *Proc., 12th World Conference on Transportation Research*, Lisbon, Portugal, 2010.
30. Adrangi, B., R. D. Gritta, and K. Raffiee. Air Cargo Shipments and Regional Employment: The Northern Nevada Case. *Journal of Business & Economics Research*, Vol. 5, No. 2, 2011, pp. 27–44.
31. Jiang, H., L. Ren, and R. Hansman. *Market and Infrastructure Analysis of Future Air Cargo Demand in China*. http://hdl.handle.net/1721.1/35819. Accessed April 2, 2017.
32. Gardiner, J., I. Humphreys, and S. Ison. Freighter Operators' Choice of Airport: A Three-Stage Process. *Transport Reviews*, Vol. 25, No. 1, 2005, pp. 85–102. https://doi.org/10.1080/0144164042000218409.
33. Kauffman, C. *Yusen Targets Intra-Asian E-Commerce with Hong Kong Cool Center*. Air Cargo World, Seattle, WA, 2016.
34. Yan, H., and H. Yuan. *Can Airport Industrial Clusters Promote the Economy?* https://doi.org/10.1109/TMEE.2011.6199133.
35. Bureau of Transportation Statistics. *T-100 Domestic Market (All Carriers, 2015)*. https://www.transtats.bts.gov/Fields.asp?Table_ID=310. Accessed October 8, 2016.
36. Boeing. *Airports with Noise and Emissions Restrictions*. http://www.boeing.com/commercial/noise/list.page. Accessed January 15, 2017.
37. FindTheData. *Commercial Real Estate*. http://commercial-real-estate.findthedata.com. Accessed December 27, 2016.
38. Top 500 Database.http://www.top500guide.com/top-500. Accessed August 2, 2016.
39. Bureau of Transportation Statistics. *Terms and Definitions*. https://www.transtats.bts.gov/DatabaseInfo.asp?DB_ID=110&DB_URL=). Accessed July 10, 2017.
40. U.S. Postal Service. *U.S. Postal Service Statement: Air Cargo Network Contract Award to FedEx*. http://about.usps.com/news/national-releases/2013/pr13_fedex-042313.htm. Accessed June 4, 2017.
41. Memphis International Airport. *Activity Report - June 2017*. 2017.
42. Hendel, D. P. *Top U.S. Postal Service Suppliers for Fiscal Year 2015*. Husch Blackwell LLP, Kansas City, MO. https://www.huschblackwell.com/~/media/files/businessin

sights/businessinsights/2016/02/postal-service-contracting-top150usps2015suppliers.pdf. Accessed March 4, 2016.
43. Goetz, A. R., and J. S. Szyliowicz. Revisiting Transportation Planning and Decision Making Theory: The Case of Denver International Airport. *Transportation Research Part A: Policy and Practice*, Vol. 31, No. 4, 1997, pp. 263–280. https://doi.org/10.1016/S0965-8564(96)00033-X.
44. Prologis. *The Global E-Commerce Revolution: What Does Its Impact on Logistics Real Estate Really Look Like?* https://www.prologis.com/logistics-industry-research/global-e-commerce-impact-logistics-real-estate. Accessed November 14, 2017.
45. Kwakkel, J. H., W. E. Walker, and V. Marchau. *Adaptive Policymaking for Airport Strategic Planning*. 2008. Accessed November 14, 2017. https://doi.org/10.1109/INFRA.2008.5439665.
46. Keogh, B. E-Retail Faces Patchwork of State Sales-Tax Laws. *The Wall Street Journal*, May 13, 2014.
47. Sales Tax Institute. *5 Things to Understand about Your Nexus Footprint*. 2016. http://www.salestaxinstitute.com/sites/default/files/Nexus_Whitepaper_2016_Update.pdf. Accessed January 5, 2018.
48. Peterson, S. *What Amazon's Sales Tax Decision Means for E-Commerce Businesses*. RetailDive, Washington, D.C., 2017.

The ACRP Selection Panel for the Graduate Research Award Program on Public-Sector Aviation Issues peer-reviewed this paper (18-03061).

Article

TRR
JOURNAL OF THE TRANSPORTATION RESEARCH BOARD

Transportation Research Record
2018, Vol. 2672(29) 59–68

DOI: 10.1177/0361198118799707
journals.sagepub.com/home/trr
SAGE

Simulation Modeling of a Prototype Designed to Address the Congestion of Passengers and Items at the Composure Area of Security Checkpoints

Maria Luisa Janer Rubio[1] **and Manuel D. Rossetti**[1]

Abstract

Discrete event simulation is used to evaluate a novel configuration of the exit roller in the composure area of security checkpoints, which is designed to address the failures of the current exit roller configuration. The paper presents in detail the conceptual modeling of a two-lane system with the current configuration, and an additional model, featuring the new design in one of the lanes. The second model is used to evaluate statistically whether there is any reduction in the passengers' system time from directing the passengers to the lane having the new design, based on the number of items passengers carry. Lastly, we perform an analysis to examine the range of arrival rates for which a single lane, featuring the new design, could replace a traditional two-lane system.

The U.S. Department of Transportation's Federal Aviation Administration has estimated that the passenger traffic for air transportation will double by 2034 (*1*). One of the major concerns is the capacity of the U.S. airports infrastructure to sustain this growth. After the terrorist attacks of September 11, 2001 (9/11), the Transportation Security Administration (TSA) required that 100% of checked baggage had to be screened. This entailed installing conveying systems and explosives detection machines across all U.S. commercial airports (*2*). Consequently, very little space was left for procedures related to passenger screening, even at current passenger numbers, without regard to possible future growth of passenger traffic. As a result, passenger screening procedures have become one of the major bottlenecks in airport operations.

TSA, as well as several research studies, has identified the x-ray procedure as the major cause of constraint to passenger flow. A study performed at Dallas/Fort Worth International Airport (DFW) attributes this to the divestiture and composure processes, happening before and after the x-ray, respectively (*3*). Divestiture represents an issue because passengers wait until they have reached the divestiture tables to start preparing for screening. In addition, once they have reached the tables, they often fail to remove certain required items from their luggage or person. At the other end of the operation, composure is an issue, because bags are delivered in the same order in which passengers divested them despite what may have happened to the passengers during screening. When, for instance, a passenger is delayed by a second security officer, the rest of the passengers continue processing. As a result, the bags of the passenger who was delayed may be blocking the bags of the following passengers, who have finished screening before the passenger who was delayed. In addition, passengers fail to move to the designated benches to put their items back in their carry-on luggage. Instead, they remain in the exit roller area. Both issues contribute to interrupting the x-ray operation, because sometimes there is not enough space on the exit roller to continue delivering items to the cleared passengers (*3*).

TSA and several airlines have taken several measures to avoid some of these issues. Divestiture officers have been employed at commercial airports to instruct passengers during the divestiture stage. Moreover, additional officers are employed during peak seasons of the year at major hub airports to urge passengers to start preparing for screening while they are waiting in line. In addition, in May 2016, Delta Air Lines introduced a pair of automated lanes at Hartsfield-Jackson Atlanta International Airport (ATL). These enable parallel processing of five

[1]University of Arkansas, Fayetteville, AR

Corresponding Author:
Address correspondence to Maria Luisa Janer Rubio: mjanerru@uark.edu

passengers during divestiture. Moreover, an automated bin system returns empty trays to passengers divesting. This has been widely used in Europe and Canada, and has been introduced in several major airports in the U.S.A. (*4*). Similarly, the Pre-Check program, established to improve the experience and security benefit of known travelers, had enrolled over 1.6 million individuals as of September 2015. TSA estimated that the number could rise to 25 million by 2020 (*5*).

Despite these multiple improvements, long lines of waiting passengers remain a concern, especially for peak seasons of the year. Over the summer of 2016, passengers experienced waiting lines of up to 50 minutes in major airports including DFW, San Francisco International Airport (SFO) and Los Angeles International Airport (LAX). The objective of this study is to address specifically the issue observed at composure where some passengers block the way of other passengers if they are delayed after body screening. We investigate the possibility of implementing a continuously circulating conveyor in place of the exit roller.

The paper is organized as follows. The following section presents a review of some of the efforts found in the literature associated with improving the performance at security checkpoints using discrete event simulation (DES) or analytical techniques. The third section describes in detail the system under study, and the prototype design. The fourth section illustrates the conceptual modeling of the simulation models. The fifth section explains how the parameters for the models' distributions were obtained. The final section describes the experimental design and summarizes the results.

Literature Review

We reviewed studies addressing the performance of airport security checkpoints based on both security and effective passenger flow. De Barros and Tomber investigated the effects of the security measures implemented after 9/11 (*6*). They used deterministic queuing theory to estimate the queue length and approximate the average waiting time at passenger screening. In addition, they used a simulation model built in Arena to evaluate seven scenarios associated with the different measures implemented after 9/11. They found that reducing the carry-on items to one item per passenger was the most effective measure to reduce the waiting times of passengers. They also discussed the advantages and disadvantages of analytical models versus DES models. They concluded that DES can be very useful when a great level of detail is needed. On the other hand, in the absence of data, analytic models can provide meaningful insights regarding the interrelationships between variables. Similarly, Dorton and Liu found queuing models are useful when the study is subject to steady-state conditions (*7*). They used queuing theory and DES to investigate the effect of baggage volume (i.e., number of items per passenger) and alarm rate on system time and passenger throughput. They concluded that alarm rate greatly affects both the system time of passengers and passenger throughput, while baggage volume has a slight effect on system time and has no effect on throughput.

Doran, Gokhale, and Lownes used an analytical model to investigate the effect of service time variability, different passenger profiles, and new policy configurations (*8*). They used hyper-exponential distributions to model service times exhibiting coefficient of variation larger than one. They modeled the various passenger profiles by decreasing and increasing the means of the hyper-exponential distributions in every service component by a certain percentage. Wetter, Lipphardt, and Hofer performed an exploratory analysis on the influence of outside temperature and the number of manual inspections on the average processing time of passengers (*9*). In addition, they studied two types of docks, one featuring a "traditional" layout having a single lane and a single walk-through medal detector (WTMD) and the other featuring a "mirror" layout with two lanes sharing one WTMD. They found the "mirror" layout provided more flexibility for the WTMD security officers to switch actively between the WTMD and as a second position for manual baggage checks. Also, the layout required less space and a single WTMD. In contrast to Dorton and Lui, they found throughput was highly sensitive to outside temperature and manual baggage inspection (*9*). Additionally, they found that taking away the divestiture officer and assigning them as a second additional person for manual baggage checks increased the number of manual baggage checks and the frequency of metal alarms from the WTMD. This intensified the work strain for security officers, but, surprisingly, it increased throughput by around 18%.

Paul et al. developed a regression model to predict the passenger waiting time based on the number of open lanes, the average throughput, and the number of items per passenger (*10*). They used a simulation model and a factorial design of experiments to determine the coefficients of the regression model. Unlike most of the studies reviewed, this study provides details about the extensive on-site data collection. From exploratory analysis on the data, they observed the highest number of items per passenger in the morning hours during weekdays, when most travelers are business passengers. In addition, the results of the simulation model showed that the factors with highest impact on waiting time are: passenger volume, number of items per passenger, number of lanes open, and staff level.

Other studies have been designed to aid the simulation analysis of security screening checkpoint systems. Guru and Savory developed an internet-based application consisting of 15 different templates to help the simulation analyst define the appropriate input model parameters, system components, and interrelationships between the passenger, the personnel, and the security equipment (*11*). Other studies present simulation templates developed to evaluate the impact of new procedures on space allocation and waiting time (*12–14*).

Similar to the present study, other efforts focus on proposing and evaluating the effectiveness of solutions to issues already identified. Leone and Liu observed that while 80% of items are inspected within seven seconds in the primary screening, a lengthy right tail in the distribution of the inspection time implies that a very small proportion of items contribute disproportionately to slowing the inspection rate and impairing the overall checkpoint performance (*15*). They proposed imposing a limit in the primary screening inspection time. Using DES, they asserted that their solution decreases mean waiting time by 43%, reduces operation costs by 1%, and increases the probability of detecting a prohibited item to 10%.

De Lange et al. explored a virtual queuing solution offering eligible passengers a time window during which they could bypass the travel document checker (TDC) queue, and have priority access to the security lanes (*16*). They evaluated the solution using DES to investigate whether virtual queuing could reduce the number of security officers while not increasing the average passenger waiting time. They found this solution works best for airports with arrival rates having sharp and frequent peaks exceeding the checkpoint capacity, followed by periods of light passenger traffic. They also found that the effectiveness of this solution is contingent upon the participation of at least 60% of the eligible passengers.

Building on the effectiveness and wide usage of simulation methods, this paper uses DES models to evaluate the effectiveness of methods to improve the composure process at security checkpoints.

System Definition

Our system consists of the passenger security screening checkpoint, including passengers, security officers, material handling, and security equipment. It extends from where passengers join the queue to have their documents checked by a TDC officer to the point where they have removed the last item from the exit roller. Figure 1 displays a layout of the system elements under study, encircled in red.

As advised by a TSA design agent, we model a pair of mirror lanes sharing a WTMD, an advanced imaging technology (AIT) scanner and a secondary screening station. Note we do not include in the system the Pre-Check lane or the processing of Pre-Check passengers. The system is modeled according to the standards in the Revision 5.1 of the TSA Checkpoint Guide (2014).

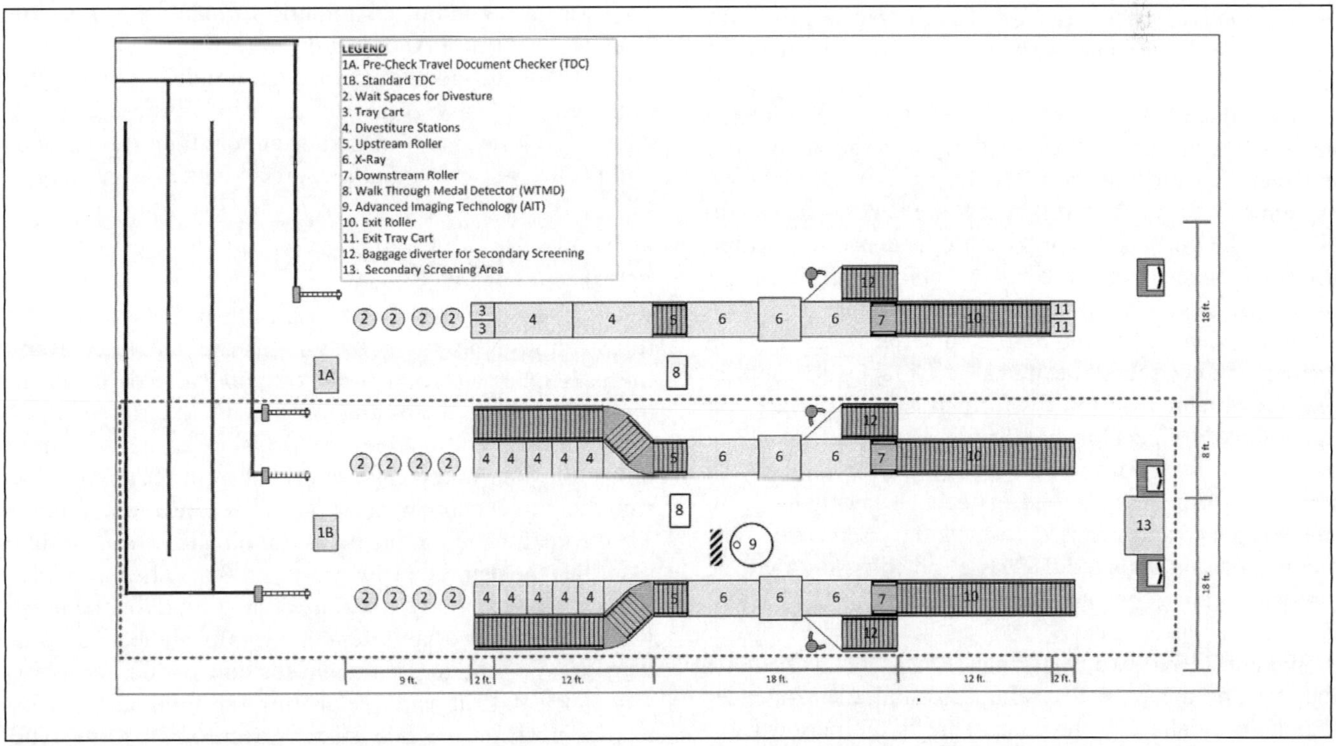

Figure 1. Security checkpoint system layout.

Upon arrival, standard passengers join the TDC queue. After the TDC officer finishes reviewing a passenger's documents, another security officer directs the passenger to the lane with the smallest number of passengers waiting. There can be at most four passengers per lane waiting for divestiture. Thus, if there is not a space available in any of the two lanes, the passenger stays with the directing officer until a space becomes available in either of the two lanes. Similarly, there can be at most two passengers waiting with the directing officer. Thus, when the system reaches this limit, the TDC officer stops processing passengers until space becomes available.

When authorized to do so, passengers proceed to divestiture. Similar to what has been recently implemented in various airports across the U.S.A., for divestiture, passengers proceed to one of five stations. Here they pick up their trays from below the counter, load one tray at a time, and push the loaded items as well as their carry-on bags to a roller conveyor that passes the items in front of an x-ray machine. Based on the findings of Paul et al., the distribution of the number of trays per passenger varies depending on the hour of the day (*10*).

The passengers' items continue through the x-ray system, where items are processed individually while the passengers proceed to the AIT station. If a suspicious item is detected by the x-ray, the tray or carry-on item is automatically diverted to secondary screening. Similarly, if a suspicious event is detected by the AIT, the suspect passenger is screened by a second officer with a wand, while the following passenger continues through the AIT. The AIT station becomes available after the security officer in charge of secondary body screening becomes available.

After the x-ray screening, cleared items are delivered to the exit roller by an 8-foot conveyor in the same order in which they entered the x-ray. Up to six trays or carry-on items fit on the exit roller. If there is not space available for the x-ray operator to continue delivering cleared trays, the operator stops the x-ray operation until a space opens for a tray to be delivered. Items that set off an alarm are diverted to the manual diverter roller. Here, a security officer takes the items individually to the secondary screening station for further screening. Carry-on items have a higher probability of setting-off an alarm than individual trays. The security officer advises the passenger to collect the rest of their cleared items from the exit roller and proceed to secondary screening. The security officer waits for the passenger to ensure that the passenger can see what the officer does with the passenger's items.

We considered two performance measures to evaluate the system: the time to composure and the passenger throughput. The time to composure is the period from when a passenger joins the TDC queue to the time when they are able to pick up their items from the exit roller. The TSA has a standard limit of 10 minutes for this measure. The passenger throughput corresponds to the average number of passengers screened by the AIT, provided that the system operates at full capacity.

Solution Prototype

The solution prototype consists of replacing the exit roller of one of the two lanes with a continuously circulating conveyor, where items circulate until passengers collect them. Figure 2 displays a layout for the solution prototype.

Up to six passengers can be waiting around the circulating conveyor to collect their items comfortably. Also, we assume passengers do not run after their circulating items. Instead, they proceed to one of the waiting spaces around the conveyor, and wait for their items to circulate in front of them. After a passenger has collected their last item, the waiting space becomes available.

The circulating conveyor is designed to occupy the same space in length as a standard exit roller, but its width occupies twice the width of an exit roller, as depicted in Figure 2.

Model Implementation

In this section, we explore the key modeling issues to represent the current configuration and the solution prototype within a computer simulation model. We used the Rockwell Software Arena environment, professional version 15. We also identified four key modeling issues: the divestiture of passengers for the models, the different composure processes for the current configuration model and prototype model, and secondary luggage screening.

Divestiture Modeling

After the passenger is able to seize one of the five divestiture stations, the passenger's items are separated from the passenger. However, first, the model needs to determine which of the items are going to be routed to secondary screening. This enables the model to have a record of how many items need to be collected from the composure roller and how many items will be at secondary screening. Therefore, for each of the items, looping logic is used that identifies the item as a tray or as carry-on. Then, it predetermines an x-ray screening processing time for the item. If this random variable is above a threshold limit, the item will be sent to secondary screening after the x-ray screening, and the variable storing the number of items to pick up from the exit roller is decreased by one. The threshold used for validating all of the models was

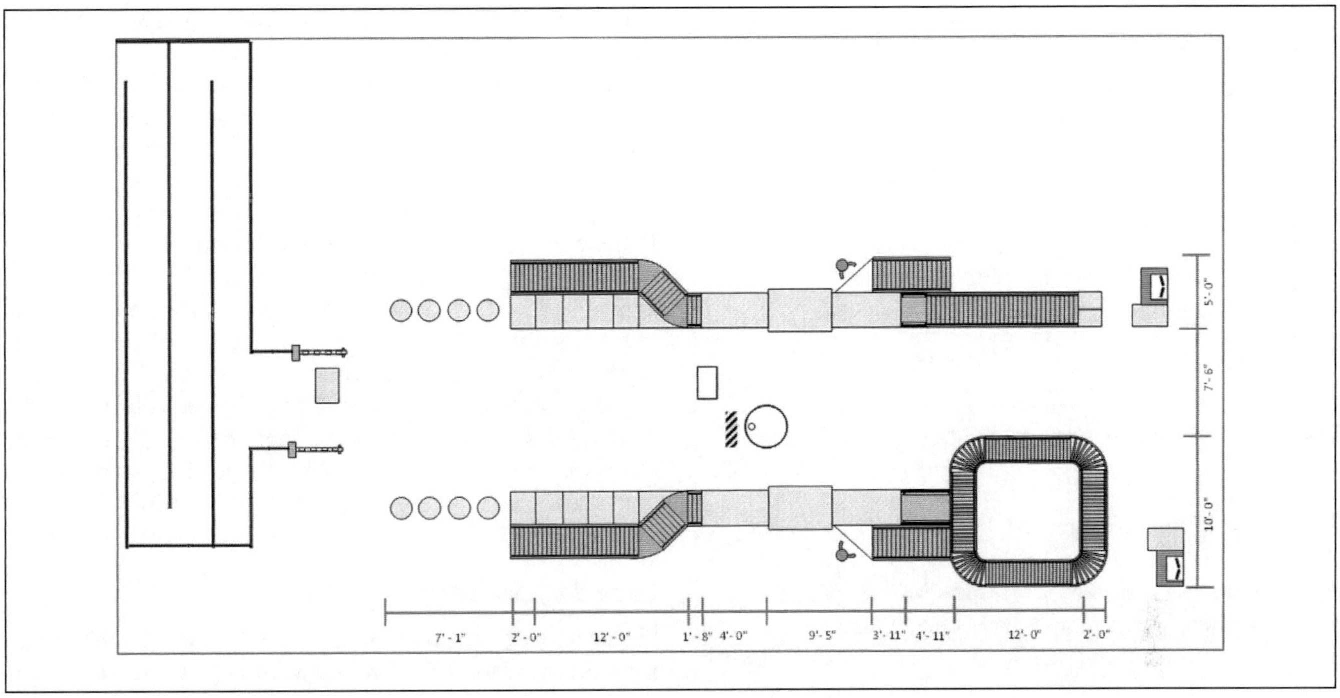

Figure 2. Prototype solution layout.

```
1 myCounter = 1
2 myCarry-On = TRUE
3 myNumItemsToPickUp = myNumItems
4 WHILE myCounter <= myNumItems
5     If (myCellIndex < myNumItems)
6         myCarry-On = FALSE
7     EndIf
8     If (myXRayPT > myXRayLimit)
9         mySecondaryScreening = TRUE
10        myNumItemsToPickUp = myNumItemsToPickUp -1
11    EndIf
12    myCounter = myCounter +1
13    SEPARATE 1 item from passenger
14 END WHILE
```

Figure 3. Logic for separating passengers from items.

11 seconds, as suggested by Leone and Liu (*15*). Figure 3 displays this logic in the while loop from Line 4 to Line 14.

After each item is created (see Line 13 in Figure 3), one by one, the items seize the space where the passenger loads each item onto the conveyor. The passenger, on the other hand, waits in a hold construct until all of their items are on the divestiture conveyor. Consequently, immediately after an item accesses the conveyor construct, it releases the loading space for the following item, and it creates a logical entity that is batched along with the rest of the passenger's entities. The logical entities represent the number of items that were divested. The final batch releases the passenger from the hold construct, notifying the passenger that all of the items are on the conveyor.

Composure Modeling—Current Configuration Model

The composure process of the current configuration was modeled using a signal-seize-hold-release logic. Essentially, it is necessary to model how items roll over the exit roller, and how they are collected by the passengers. Figure 4 illustrates the modeling of this logic.

First, the item being delivered signals the item in the hold of the first space, so this item moves to the next space. The signaling item then seizes the first space and exits the composure conveyor. If there is no item coming immediately behind it, the item enters a hold construct and waits for the signal of a rolling item or for the passenger to pick it up. This is the same for spaces 2 through 4. If an item rolls all the way until space 5 of the exit roller, the item enters an infinite hold, where it waits until its passenger picks it up.

In the first space, every item creates a logical entity that is sent to match with the passenger. This entity acts like a messenger which notifies the passenger that one of their items is ready to be picked up. After the passenger finds a matching entity in the match construct, the passenger proceeds to look for the item in question.

A loop construct is used to search for the items. A counter is initialized to 5, so the passenger starts searching from space 5 (i.e., the farthest away from the x-ray),

```
SIGNAL item in Hold S1
SEIZE Resource S1
EXIT Composure Conveyor
SEPARATE dummy item
// send dummy to MATCH with passenger
WHILE myCounter <= 4
        If(NQ(MEMBER(QueueSpacesSet, myCounter)) <= 0)
                WAIT for signal(myCounter)
        End if
        myCounter = myCounter + 1
        SIGNAL item in HoldSpaceSet(myCounter)
        SEIZE ResourceSpaceSet(myCounter)
        RELEASE ResourceSpaceSet(myCounter - 1)
END WHILE
Infinite HOLD for S5
```

Figure 4. Logic for item rolling on exit roller.

and the counter is decreased by one as the passenger moves closer to the x-ray. In this way, we ensure the passenger collects the item corresponding to the logical entity to which they are matched. The passenger searches the resource queue and the hold queue associated with every space. If the passenger finds their item in a hold queue, the passenger removes the item from the queue, and the item releases the associated space associated because this implies that the item is physically in that space. However, if the queue is a resource queue, this implies that the item is in the queue to seize the space associated with the current space, but the item is physically in the previous space. Thus, upon being removed from the queue, the item releases the space associated with the counter minus one.

After the passenger removes an item, the passenger checks whether additional items need to be collected. If this is true, the passenger returns to the match construct to be matched with another dummy entity that represents the items to be picked up and repeats the search process again. If the passenger has collected all the items that need to be picked up from the roller, but the passenger does not have all of their initial items, the passenger goes to the secondary screening station. Otherwise, the passenger leaves the checkpoint.

Composure Modeling—Circulating Conveyor

The composure process for the prototype model follows a different logic than in the current configuration for the lanes where the exit roller is replaced by a continuously circulating conveyor.

If the passenger needs to pick up some items from the circulating conveyor, the passenger seizes the closest space available around the conveyor. The passenger delays some random time for walking to the space, and enters a hold construct linked to the space that was seized. Items are delivered to the circulating conveyor by the composure conveyor. Upon accessing the circulating conveyor, the items exit the composure conveyor and convey by sequence from space 1 to space 6 of the conveyor. In each space, items are assigned an index associated with the space at which they are located. The active entity (the item) searches for the passenger in the hold queue of the space associated with their current station. If the entity (item) finds a passenger with the same serial number in a particular space, the item exits the conveyor and is batched with the other items of the passenger. If items do not find a passenger, they continue circulating to the next station. When the item reaches station 6, and has not been batched with its passenger, the attribute *Entity.JobStep* is reset to 0, so that the entity (item) can circulate again from station 1 of the composure conveyor.

Secondary Screening

Similar to divestiture, secondary screening is the same for both models. After the items are conveyed out from the x-ray, they are directed either to the composure conveyor or to the diverter roller (if they require secondary screening). The diverter roller is inclined, so items roll to the end of the roller with the force of gravity. Thus, the diverter roller was also modeled using a conveyor construct. Figure 5 displays the flow logic of items on the diverter roller.

Upon accessing the diverter roller, items exit the downstream roller. Next, they are conveyed to the end of the roller, and the items of a single passenger are grouped together as a representative entity. This entity seizes the secondary security officer. Next, the representative entity is separated again into individual items, and each item creates a duplicate. The duplicates proceed to a match construct while the actual items are grouped again into a representative entity. This entity is a representation of the officer itself, who goes back and forth between the diverter roller and the secondary screening station to bring the items of the passenger to the secondary screening station.

Thus, in the diverter roller, this officer entity is matched with one of the duplicates so the duplicate exits the conveyor while the officer entity walks to the secondary screening station. This logic was necessary so items would only exit the diverter roller after the officer picked them up to take them to secondary screening. The officer entity goes back to the diverter roller and repeats the process until the officer has taken all the items of the passenger to the secondary screening station. Then, the secondary screening officer is released to serve other passengers while the items' representative entity waits to be matched with the passenger. Upon passenger arrival, the secondary screening officer is seized again, so the

```
If (mySecondaryScreening == FALSE)
        GOTO composure
Else
        ACCESS diverter roller
        RELEASE downstream roller
        CONVEY to the end of diverter roller
        BATCH items of passenger temporarily (Entity.SerialNumber)
        SEIZE 2Screening Officer
        SEPARATE temporary batch
        SEPARATE one dummy per item
```

Figure 5. Logic for item on diverter roller.

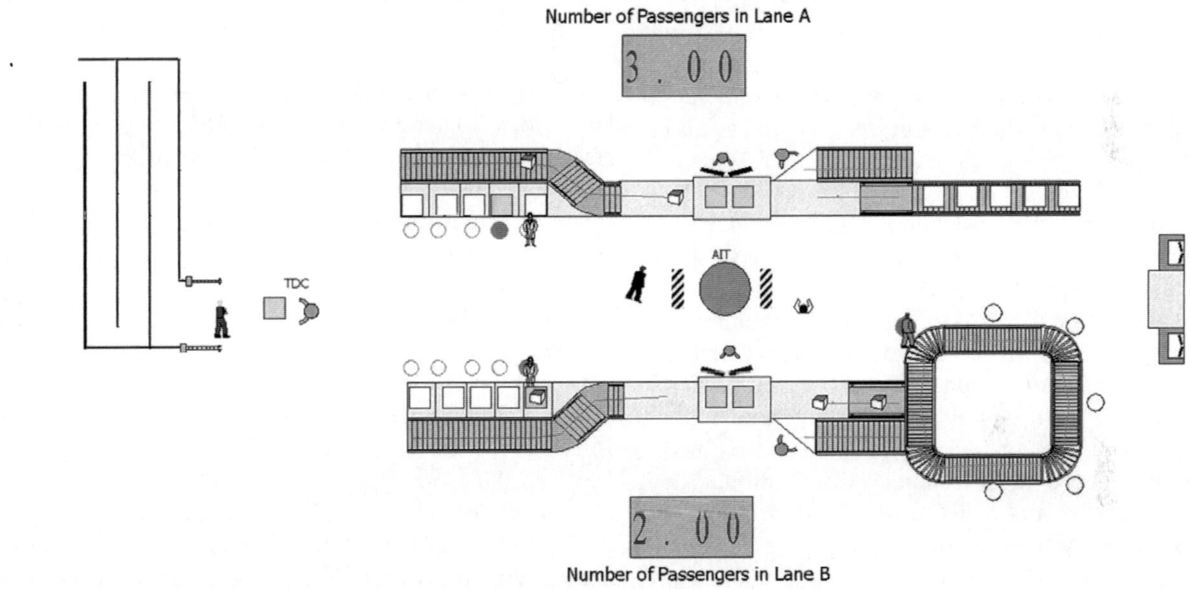

Figure 6. Animation of solution prototype.

secondary screening can take place. After the secondary screening, a logical entity is created to delay for the time that the secondary officer takes to walk to the diverter roller while the items and passenger leave the system.

Model Verification and Validation

For verifying the models, we carefully animated the key modeling issues in the models, while debugging what seemed strange or unusual behavior. Figure 6 displays a picture of the animation of the solution prototype. This presents the difference between the animation of a traditional lane and a circulating conveyor lane.

For validating the models, we compared the current configuration model to the observations and findings of other studies found in literature. According to the airport classification in Leone and Liu, the volume of passengers taken into account for this study closely mimics the volume of a medium hub airport where peak-hour passenger arrival volumes reach 200 passengers per hour (pph) for a two-lane system (*15*). According to their literature review, waiting times at medium hub airports are slightly below the TSA standard 10-minute limit, and the AIT runs at a capacity of 150 passengers per hour (*15*). Using an arrival rate of 200 passengers per hour, a threshold of 11 seconds for the x-ray processing time, as suggested by Leone and Liu's study, we obtained an average AIT throughput of 165.71 passengers per hour with a half width of 25.40 passengers based on a 95% confidence interval. In addition, we obtained an average waiting time of 3.56 minutes with a half width of 0.34 minutes. Also, an average 91% of the passengers with a 0.02 half width have a waiting time under 10 minutes. Note that our system seems to have slightly better performance measures than the system studied by Leone and Liu, because our system accounts for the new automated divestiture system implemented at ATL in 2016. This system is expected to reduce waiting times by 30% (*4*).

Distribution Parameters

Given the security restrictions associated with collecting live observations at security checkpoints, we obtained the parameters for the models from data collection efforts found in the literature. We used a mean arrival rate of 200 pph, because the system in this study was classified as a checkpoint of a medium hub airport (*15*). Paul et al. provide three discrete distributions for the number of items per passenger for high, medium, and low passenger traffic volumes (*10*). Moreover, we approximated the distributions for loading items during divestiture and unloading items during composure using triangular distributions. For the loading of items, we used parameters of 2, 7, and 10 seconds per carry-on item, and 1, 4, and 7 seconds per tray. For the unloading of items, we used parameters of 2, 7 and 15 seconds for all items. For the x-ray inspection time, we used Leone and Liu's gamma distribution with shape and scale of 3.49 and 2.05 seconds, respectively (*15*). Items requiring inspection times above 11 seconds were directed to secondary screening as in the models presented by Leone and Liu (*15*). For the secondary screening distribution, we used the gamma distribution obtained from Nie et al. with shape and scale of 2.00 and 2.05 minutes respectively (*17*). Lastly, Hobbs, Rossetti, and Faas provide a triangular distribution with parameters 2.93, 4.40, and 5.86 feet per second for a normal human walking velocity (*18*).

Experimentation and Results

The experimentation for this study consisted of three parts. First, we compare the performance of the traditional exit roller against the new circulating conveyor design. Then, we investigate the range of arrival rates under which a single circulating conveyor lane could replace a traditional two-lane system. Lastly, we use the prototype model to examine whether processing passengers with the largest number of items in the circulating conveyor lane could result in any reduction to the passengers' system time.

For the first and second parts of the experimentation, we used three different versions of the simulation models: the base model, featuring two traditional lanes; a second model, featuring a single circulating conveyor lane; and a third model, featuring a single traditional lane. For the three models, we evaluated the response of three performance measures to the change in arrival rates. The performance measures included the mean passenger throughput, the mean system time, and the probability that the time to composure is less than 10 minutes. We used eight different arrival rates ranging from 150 pph to 200 pph. For each arrival rate, we obtained 35 observations of the three responses, and analyzed them statistically.

Using the single traditional lane model and the single circulating conveyor lane model, we found that at 150 pph, the probability that the time to composure is less than 10 minutes is 100% for the circulating conveyor lane model, while it is just over 30% for the single traditional lane model. Similarly, the model featuring the circulating conveyor lane results in an AIT throughput of 150 pph, meaning that all the arriving passengers in an hour are screened within the same hour. On the other hand, the single traditional lane model processes only 122 pph. Moreover, the mean system time for the single circulating conveyor lane model is a little under 3 minutes while it is over 20 minutes for the single traditional lane model. Therefore, the circulating conveyor design clearly outperforms the traditional exit roller.

In light of these results, we decided to investigate the range of arrival rates for which a single circulating conveyor lane can achieve performance equivalent to that of a traditional two-lane system. We found that the estimated probability that the time to composure is under 10 minutes for the model featuring a single circulating conveyor lane was statistically the same as that of the traditional two-lane system only with arrival rates under 150 pph (p-value = 0.083). For an arrival rate of 180 pph, the estimated probability for the single circulating conveyor lane was 0.723, statistically lower than that of the two traditional lanes system, which had an estimated probability of 1. Nonetheless, the mean passenger throughput at this arrival rate was statistically the same for both models (p-value = 0.143). In addition, with the same arrival rate, the mean system time for the single circulating conveyor was 8.09 minutes, significantly lower than 10 minutes, the TSA standard for the time to composure. Consequently, for an arrival rate of 180 pph, the TSA could replace the traditional two-lane system with a single circulating conveyor if they are willing to tolerate 70% as the estimate for the probability that the time to composure is at most 10 minutes. Thus, any additional space requirements associated with the wider exit roller are mitigated by the fact that a single conveyor can meet the performance of two conveyors in the proposed system.

For the second part of the experimentation, we designed a different rule for directing passengers to the lanes in the prototype model. In the TDC process, the directing officer sends passengers, having a number of items below a specific threshold, to the lane having the least number of passengers in the divestiture area, as in the original model. On the other hand, passengers having a number of items equal to or greater than the threshold will be directed to the circulation conveyor lane if there is space for them to wait in one of the four spaces before the divestiture stations. Otherwise, they will be sent to the lane having the least number of passengers in the divestiture area.

We used the three distributions on the number of items per passengers, obtained from Paul et al. (*10*). In any of the three distributions, passengers may have one to five items or zero items. Thus, we varied the threshold on the number of items from one to six items. We did not include a threshold of zero items, because the system time of a passenger with no items would be essentially the same regardless of the lane in which they are processed. The six-item threshold corresponds to the initial configuration of the prototype, where passengers are directed to the lane having the least number of passengers in the divestiture area.

For each threshold, we obtained 30 replicates of the system time. Then, we performed a multiple comparison of the mean system time under the six different thresholds for each of the three distributions. For the high and medium traffic-volume distributions, we found that the system time was statistically lower for thresholds on four, five, and six items. However, the difference among the system times of the three thresholds was not significant. Similarly, for the low traffic-volume distribution, the system time was statistically lower for thresholds on three, four, five, and six items. Nonetheless, the difference between the system times of these thresholds was not significant either. Therefore, setting a threshold to direct passengers to the circulating conveyor lane in the prototype model, such as passengers with more items being processed in the circulating conveyor lane, does not reduce the passengers' system time significantly.

Future work of this study should examine how the circulating conveyor affects the x-ray and secondary screening processes. In particular, we should examine how we could leverage on the effect of the circulating conveyor on the system time to decrease the risk associated with ignoring a potential threat. Because the conveyors may change the order of output of bags from the x-ray process, future research will need to examine the effect of the new divestiture conveyor and the proposed composure conveyor on the ability of passengers and screeners to match the correct bag with the correct passenger.

Author Contributions

The authors confirm contribution to the paper as follows: study conception and design: MLJR, MR; data collection: MLJR, MR; analysis and interpretation of results: MLJR, MR; draft manuscript preparation: MLJR, MR. All authors reviewed the results and approved the final version of the manuscript.

References

1. Price, H. *FAA Forecast Sees Continued, Steady Growth in Air Travel.* Federal Aviation Administration, Washington, D.C., 2015. https://www.faa.gov/news/press_releases/news_story.cfm?newsId = 18434. Accessed January 17, 2017.
2. Blalock, G., V. Kadiyali, and D. H. Simon. The Impact of Post-9/11 Airport Security Measures on the Demand for Air Travel. *Journal of Law and Economics*, Vol. 50, No. 4, 2007, pp. 731–755.
3. DFW Planning Department. *Dallas/Fort Worth Security Checkpoints - Tiger Team 2005 - Improving Throughput.* 2006. https://www.dfwairport.com/dfwucm1prd/groups/public/documents/webasset/p1_008154.pdf. Accessed January 17, 2017.
4. Solomon, R. *TSA Chief: Delta Innovation Lanes Make "Dramatic Improvements."* 2016. http://news.delta.com/tsa-chief-delta-innovation-lanes-make-dramatic-improvements. Accessed January 17, 2017.
5. TSA Pre-Check Expansion Act. Report of the Committee on Commerce, Science, and Transportation on H. R. 2843. 2016. https://www.congress.gov/bill/114th-congress/house-bill/2843/.
6. De Barros, A. G., and D. D. Tomber. Quantitative Analysis of Passenger and Baggage Security Screening at Airports. *Journal of Advanced Transportation*, Vol. 41, No. 2, 2007, pp. 171–193.
7. Dorton, S. L., and D. Liu. Effects of Baggage Volume and Alarm Rate on Airport Security Screening Checkpoint Efficiency using Queuing Networks and Discrete Event Simulation. *Human Factors and Ergonomics in Manufacturing*, Vol. 26, No. 1, 2016, pp. 95–109.
8. Doran, D., S. Gokhale, and N. Lownes. Analytic Model of Screening Times at Airport Security Checkpoints. *Transportation Research Record: Journal of the Transportation Research Board*, 2013. 2400. 1–8.
9. Wetter, O. E., M. Lipphardt, and F. Hofer. External and Internal Influences on the Security Control Process at Airports. *Proc., IEEE International Carnahan Conference on Security Technology*, San Jose, CA, IEEE, New York, NY, 2010, pp. 301–309.
10. Paul, J. A., R. Batta, L. Lin, and C. Drury. Airport Security System Design: Passenger Flow Analysis and Simulation Modeling. In *Protecting Airline Passengers in the Age of Terrorism* (Seidenstat, P., and F. X. Splane, eds.), ABC-Clio, Santa Barbara, CA, 2009.
11. Guru, A., and P. Savory. A Template-Based Conceptual Modeling Infrastructure for Simulation of Physical Security Systems. *Proc., 2004 Winter Simulation Conference*, Washington, D.C., IEEE, New York, NY, 2004, pp. 866–873.
12. Fayez, M. S., A. Kaylani, D. Cope, N. Rychlik, and M. Mollaghasemi. Managing Airport Operations using Simulation. *Journal of Simulation*, Vol. 2, No.1, 2008, pp. 41–52.
13. Wilson, D., E. K. Roe, and S. A. So. Security Checkpoint Optimizer (SCO): An Application for Simulating the Operations of Airport Security Checkpoints. *Proc., 2006 Winter Simulation Conference*, Monterey, CA, IEEE, New York, NY, 2006, pp. 529–535.

14. Pendergraft, D. R., C. V. Robertson, and S. Shrader. Simulation of an Airport Passenger Security System. *Proc., 2004 Winter Simulation Conference*, Washington, D.C., IEEE, New York, NY, 2004, pp. 874–878.
15. Leone, K., and R. Liu. Improving Airport Security Screening Checkpoint Operations in the US via Paced System Design. *Journal of Air Transport Management*, Vol. 17, No. 2, 2011, pp. 62–67.
16. De Lange, R., I. Samoilovich, and B. V. Der Rhee. Virtual Queuing at Airport Security Lanes. *European Journal of Operational Research*, Vol. 225, No. 1, 2013, pp. 153–165.
17. Nie, X., G. Parab, R. Batta, and L. Lin. Simulation-Based Selectee Lane Queueing Design for Passenger Checkpoint Screening. *European Journal of Operational Research*, Vol. 219, No. 1, 2012, pp. 146–155.
18. Hobbs, B., M. D. Rossetti, and P. Faas. An Object-Oriented Framework for Simulating Automatic Data Collection Systems. *Proc., 2006 Winter Simulation Conference*, Monterey, CA, IEEE, New York, NY, 2006, pp. 1545–1553.

The ACRP Selection Panel for the Graduate Research Award Program on Public-Sector Aviation Issues peer-reviewed this paper (18-03066).

Article

TRR
JOURNAL OF THE TRANSPORTATION RESEARCH BOARD

Transportation Research Record
2018, Vol. 2672(29) 69–81

DOI: 10.1177/0361198118799701
journals.sagepub.com/home/trr
SAGE

Modeling Airport Business Risks, Enplanement Volatility, and Valuation of Flexibility Options in Airport Expansion Projects

Ilker Karaca[1]

Abstract

Airport capital improvement programs involve considerable flexibility in investment timing and engineering design. Even though the valuation of flexibility options may depend on several factors, the volatility of future airport activity levels, which largely defines the business risk for airport operators, makes up the focus of the present paper. As such, the paper proposes a model that can be used to value two types of flexibility options common in airport expansion projects. The first type of option—flexibility in investment timing decisions—creates value by conditioning capacity expansion decisions on trends in airport activity levels. The second, flexibility in engineering design, permits operators to influence their demand composition and to reconfigure airport facilities if their business environment changes unfavorably. A Monte Carlo simulation example also demonstrates the application of the proposed valuation model. The results show that flexibility options add economic value by reducing downside exposures and by providing the ability to increase capacity if enplanements stay on a rising trajectory. Moreover, the paper provides a comparison of enplanement growth rates by airport size for the largest 140 U.S. airports from 1990 to 2016. The analysis shows that medium airports may be uniquely positioned to benefit the most from flexible design approaches. For these airports, results imply higher exposures to excess capacity risks because of the increased persistence of losses, despite the higher year-on-year volatility of small airports.

With constantly evolving future infrastructure needs, air traffic and passenger volumes, and shifting of airline hubs, airport capital investment decisions involve unique challenges and necessitate an increasingly dynamic approach to the availability of new information. Combined with the sustained growth in air travel and freight demand, the expansion capacity of smaller airports to alleviate the mounting pressure on gateway airports is also becoming a central issue. In fact, airport expansion projects remain as the main method of adding capacity to the air transportation system. Yet expansion projects may impose considerable opportunity costs when expected demand levels either do not materialize or prove short-lived. Similarly, failure to exercise expansion capabilities in a timely fashion may lead to lost opportunities in air service growth.

As a result, a better understanding of the changes in airport activity levels is crucial in valuing capital investment projects. As such investments are justified on both the adequate and the sustained nature of the future activity volumes, significant deviations from forecasts may lead to substantial shortfalls in expected project revenues. Premature investments and non-optimal airport sizes can burden operators for years to come. The recent experiences of several U.S. airports that experienced rapid growth only to be faced with severe activity losses after losing their hub airlines exemplify the magnitude of the risks associated with major expansion programs.

In approaching these challenges, airport planners and operators can draw from a considerable amount of research that highlights the relevance of real options for investments under uncertainty, such as those considered in this study ([*1–5*], for example, offer several motivating examples and practical applications of real options in capital investment projects). In essence, investments valued within a real options framework take into account the decision makers' flexibility in deferring, abandoning,

[1]Department of Finance, Iowa State University, Ames, IA

Corresponding Author:
Address correspondence to Ilker Karaca: ikaraca@iastate.edu

and expanding projects as new information is revealed over the course of a project's life. In a similar vein, airport expansion projects arguably strive to strike a balance between creating the capacity to serve *anticipated* demand volumes and deferring expansion decisions until the demand patterns are well established. Consequently, understanding both the upside potential and the downside risks associated with expansion projects is essential to the valuation of such projects.

Thus, contributing to the growing literature (*6–9*) that recognizes the value of flexibility in engineering and airport development projects, this paper offers an explicit specification of the demand risks U.S. airport operators may face (hereafter referenced "jumps") in their enplanement volumes. The study singles out enplanement volatility as a key valuation variable for its far-reaching implications to several aspects of airport capital improvement decisions. Other additional considerations, such as airport user pricing and fees, the structure of airline use and lease agreements, cost per enplanement (CPE) rates, and competition from other airports, can also be highly relevant in the eventual implementation of capital improvement projects (*10–12*). Although the risks related to airport user pricing and fees, for instance, can be crucial in the economic viability of a project, the volatility of airport activity levels is often among the leading factors with significant potential for creating option values. As a Federal Aviation Administration (FAA) advisory circular on airport master plans states it, "passenger levels are particularly important, since they determine the size of the terminal building and other important elements of airport infrastructure such as parking facilities and access roads" (*13*).

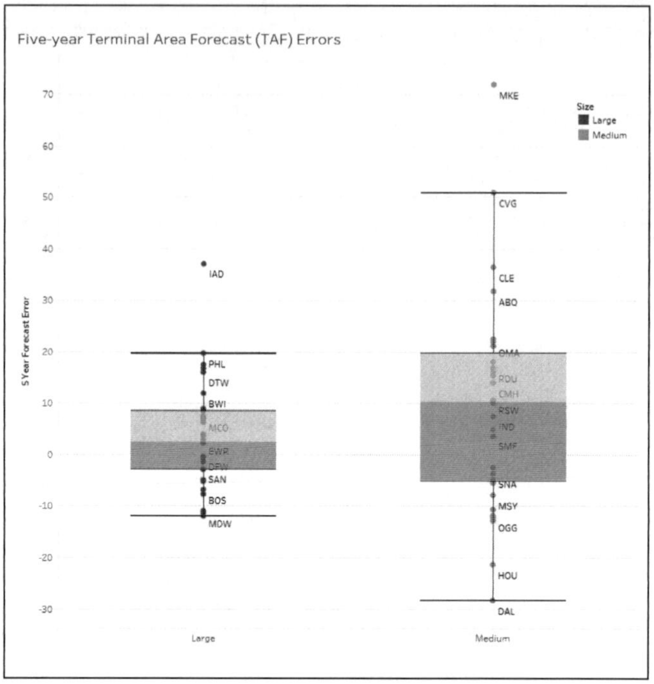

Figure 1. Forecast error of FAA passenger enplanement estimates (5-year forecasts in 2010 versus 2015 actuals). Positive values indicate lower forecast levels than actual (*18*).

Uncertainty in Airport Activity Volumes and Capacity Utilization Levels

Airport investments are subject to considerable uncertainty in airport activity volumes (*12*). Compounding the high uncertainty in activity levels are the substantial size and irreversibility of such investments. Although extensive forecasting efforts precede most airport investments, such forecasts can be highly inaccurate. Figure 1, for example, shows the magnitude of such errors for large and medium airports in the U.S., based on a sample that compared 5-year projections by the FAA against actual enplanement levels. During the 5-year sample period (2010–2015) shown in the figure, for instance, forecasts for medium airports not only overestimated the actual enplanement growth, but errors were also considerably higher for this group of airports. Yet because the long investment horizons of airport capital improvement projects also often lead to unmanageably large confidence intervals, and even though such estimates are almost always inaccurate (*14*), point estimates for activity level forecasts can quickly become focal points that drive the final investment decisions.

If forecast errors are expected to be sizable then, such variations from the mean estimates can be consequential for the operators and cannot be ignored. Fortunately for airport operators, however, airport capital investment programs may furnish the ability to time expansion investments and to accommodate engineering designs flexible enough to adapt to changing needs (*15*). As de Neufville and Scholtes emphasize, "a flexible design permits but does not require expansion" (*6*). The purpose of this paper, therefore, is to provide a modeling framework to outline the main capacity and revenue trade-offs present in airport expansion projects.

The model proposed by the paper reflects many of the practical challenges faced by airport operators in their efforts to value their expansion options. As Horonjeff et al. (*10*) point out, in the aftermath of the September 11, 2001 terror attacks, and the wave of consolidation and mergers activity following the 2008 recession, airport planning processes have become increasingly cognizant of rare events and their effect on demand forecast. For this reason, the study considers not only a trend-stationary baseline demand equation but additional uncertainties to account for similar unexpected events (see Caves and Gosling for several cases of how airport planning practices can benefit from phasing of investment decisions in the long term [*16*]).

In a recent report by the Airport Cooperative Research Program on addressing uncertainty in airport planning, Kincaid et al. provide extensive discussion of how capital investment programs can include provisions, risk mitigation strategies, and scenario planning for volatile airport activity volumes, rare and high-impact events, and political risks (*12*). Of the several risk management strategies mentioned in the ACRP report, flexibility options on common-use facilities and equipment, planning for incremental development (such as modular terminal designs that can be scaled up and down), and multi-functional designs provide the motivating flexibility options for the present paper. In particular, this study illustrates two mechanisms through which options can add incremental value to capital improvement projects. In so doing, the study provides a Monte Carlo simulation example to demonstrate how capacity levels may evolve as the future activity level paths are revealed within the model's volatility assumptions. The ACRP report also provides an outline for a multi-step approach to develop and evaluate risk response strategies for airports based on the identification and quantification of the uncertainties in airport activity levels (*12*). Following a similar approach, this paper concentrates on the assessment of the enplanement volatility for the largest U.S. airports and proceeds to the valuation of flexibility options.

For the purposes of this study, when referenced, expansion projects relate to incremental airport capacity increases (*10*, *11*), which may be modular in design. The addition of new terminal gates are examples of such projects, for instance. Although the results of the model would still be expected to apply to other capital improvement projects, such as the addition of runways, several components of the expansion model presented here would need to be altered to reflect the unique features of such projects fully.

The remainder of the paper is organized into three main sections. Part I is dedicated to the examination of the volatility of passenger enplanements and capacity utilization levels. Because the dispersion in airport activity demand is expected to be the main driver of option values, a comparative analysis of enplanement growth by airport size complements the paper's analysis. The statistical results are drawn from an examination of the enplanement growth of the largest 140 U.S. airports for the 26-year period from 1990 to 2016 (*17*). Part II provides a brief discussion of the two types of flexibility options and the presentation of the proposed simulation model. It also introduces the input parameters that are used in the modeling effort. Part III offers the simulation results for a hypothetical expansion project, which provides support for the incremental value added through adopting flexible design methods. Finally, the paper concludes with a review of the main findings.

Part I. Volatility of Enplanements and Capacity Utilization Levels for U.S. Airports

Methodology

There is an extensive body of work that employs Monte Carlo simulations in valuing real options (see [*18*] and the references therein for several examples of Monte Carlo simulations). Airport capital investments and, in particular, the role of uncertainty in flexible engineering design have also been an active research area (*7*, *11*, *12*, *16*, *19*). In line with the existing literature on real options valuation, this study follows the four-step methodology proposed by de Neufville and Scholtes (*6*) as described below:

- *Step 1.* Recognize the uncertainties in the project. Passenger enplanements are the main driver of uncertainty in this model.
- *Step 2.* Identify the types of flexibility that are the most suitable for the uncertainties recognized in Step 1.
- *Step 3.* Choose the optimal flexible design strategy and incorporate into the design.
- *Step 4.* Plan for the implementation of the flexible design strategies by monitoring the conditions deemed suitable to exercise flexibility options. (Because this step involves continuous monitoring of and the necessary adjustments to the investment strategies, it is not included in what follows.)

Step 1. Volatility in Airport Activity Levels

Patterns in Growth Rates. This section discusses the relationship between airport sizes and the volatility of enplanement growth rates for 140 leading U.S. airports (Table 1) (*17*). Table 2 provides the summary statistics for both the enplanement level and growth rate time series. Studying the rate and not the level of enplanement growth is essential for building realistic valuation models because the stationarity of the rate series can provide valuable insights into the expected enplanement growth rates over time. These expected rates (also called drift) then can be used to account for the overall steady growth in air transport.

Unlike the steady average enplanement growth forecasts issued by the FAA for U.S. passenger enplanements (*17*), airport activity levels are subject to substantial volatility and invariably differ from point estimates. Although enplanement volumes seem to be

Table 1. Summary Statistics of Passenger Enplanement Data by Airport Size (1990–2016)

	Large	Medium	Small	All
Yearly passenger enplanements				
Mean	15,508,816	3,945,670	775,726	4,624,918
Median	13,853,299	3,581,177	612,518	1,529,334
Max	50,120,617	15,395,308	3,311,931	50,120,617
Min	1,980,046	1,030,627	117	117
SD	8,150,153	1,966,017	487,669	6,917,821
Skewness	1.34	2.11	1.08	2.54
Kurtosis	4.88	9.77	4.01	10.48
Observations	783	918	2052	3753
Yearly change in enplanements				
Mean	0.0259	0.0161	0.0220	0.0214
Median	0.0264	0.0201	0.0111	0.0177
Max	0.3793	0.7552	1.6794	1.6794
Min	−0.2865	−0.4382	−0.4571	−0.4571
SD	0.0621	0.0836	0.1225	0.1035
Skewness	0.2986	0.5894	3.2268	3.0037
Kurtosis	7.2618	14.4378	31.3139	35.2390
Observations	750	884	1,956	3,590

Note: Max = maximum; Min = minimum; SD = standard deviation.

Table 2. Summary of Equality Tests for Changes in Enplanements by Airport Size

Method	Df	Test Statistic	Probability
Test for equality of means of yearly changes in enplanements			
Welch-ANOVA F-test	(2, 2086)	3.7177	0.0245
Test for equality of medians of yearly changes in enplanements			
Med. Chi-square	2	37.9716	0.0000
Kruskal-Wallis	2	25.9264	0.0000
Test for equality of variances of yearly changes in enplanements			
Bartlett	2	479.3247	0.0000
Levene	(2, 3587)	42.9518	0.0000
Brown-Forsythe	(2, 3587)	40.0354	0.0000

Note: Df = degrees of freedom.

indistinguishable among the three airport size categories at first glance (Figure 2), for instance, on closer inspection of the dispersion of growth rates, a distinct pattern emerges within the airport size groups (Figure 3).

If the volatility of activity levels is stationary, however, limiting the valuation analysis to yearly volatilities alone would lead to misleading conclusions. Because the presence of mean reversion in capacity utilization levels, for instance, would imply that the effect of severe drops would tend to dissipate over time and the airport would converge toward its mean capacity utilization level in the long run, high growth rate volatility would not necessarily lead to high option valuations. Therefore, flexibility option valuations are expected to be the greatest, especially for airports that follow a random walk process in their capacity utilization levels. As such, yearly volatility of growth rates and actual levels of enplanements can give divergent outcomes if the series for different airport sizes follow different time series processes.

A Proxy for Airport Capacity Use. This section proposes a proxy measure for changes in airport capacity utilization levels because, in addition to the standard net present value (NPV) rule, minimizing the excess capacity levels of the expanded airport operations can be a central consideration in the airport operators' capital investment decisions. Because of the emphasis placed on enplanement volumes, airport capacity refers to the enplanements alone, and not to alternative delay-based capacity estimation metrics, such as annual service volume (ASV), and Aviation System Performance Metrics (ASPM) (*21*). Therefore, Equation 1 gives airport capacity utilization levels over time by normalizing yearly enplanement levels

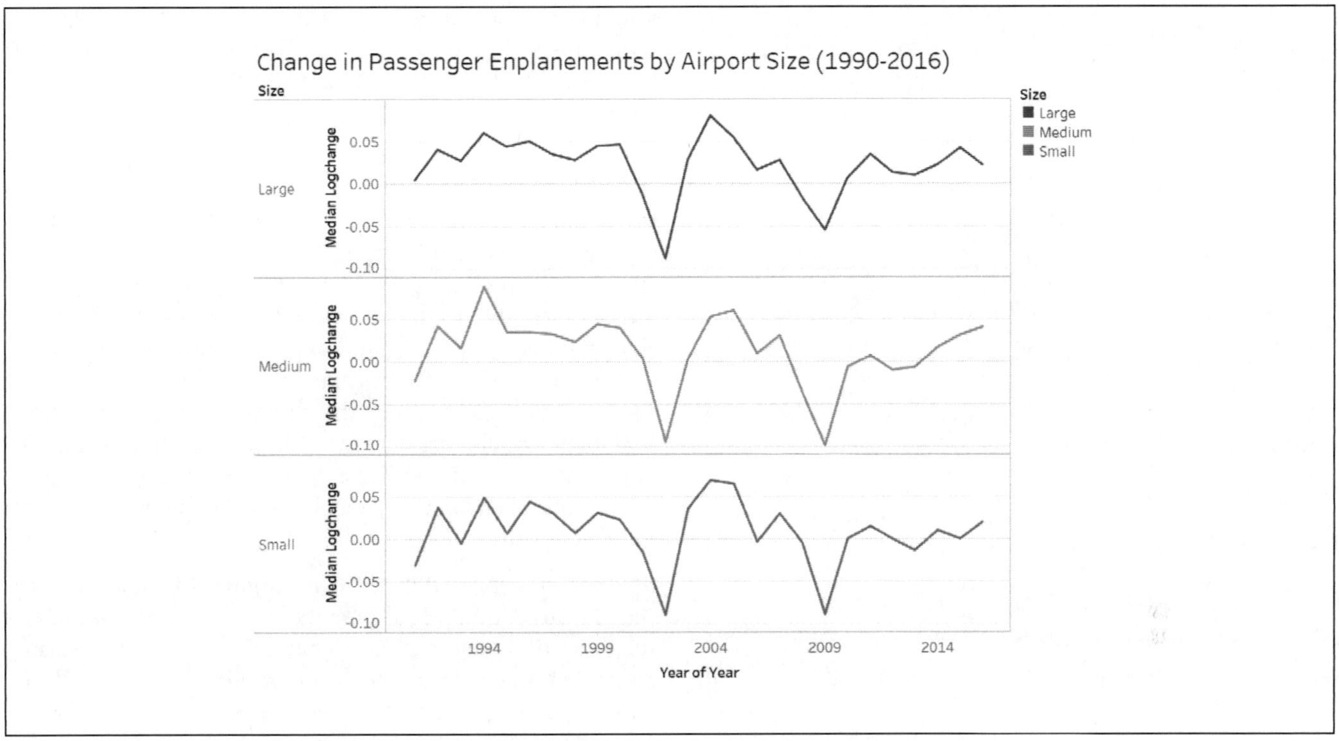

Figure 2. Change in passenger enplanement by airport size from 1990 to 2016.

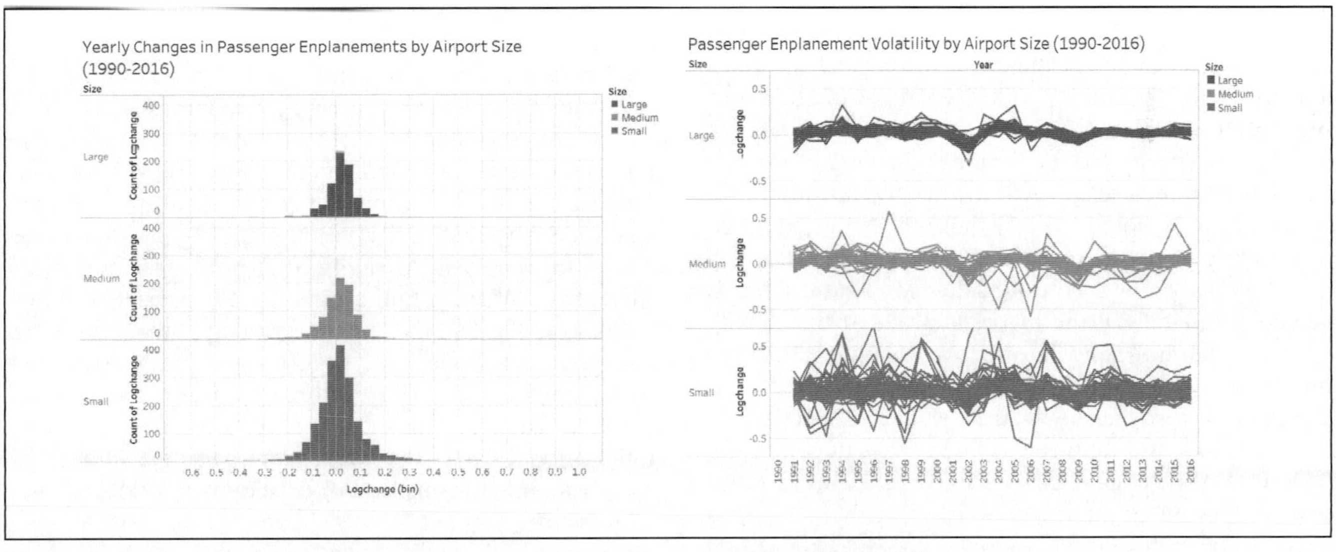

Figure 3. Distribution of changes in passenger enplanement by airport size from 1990 to 2016.

by the maximum enplanement volume previously recorded. The resulting panel data series is expected to provide sharper insights into the relative persistence of losses for medium airports when compared with their smaller and larger counterparts. Therefore, to approximate airport capacity utilization levels, a capacity utilization proxy (CUP) for the airport i in year t is defined as:

$$\text{Capacity utilization proxy}_{i,t} = \frac{\text{enplanements}_{i,t}}{\max(\text{enplanements}_{i,1990}, \ldots, \text{enplanements}_{i,t})} \quad (1)$$

As a consequence, whereas a capacity utilization of 100% for an airport indicates a record enplanement level in a year, lower values indicate the presence of excess

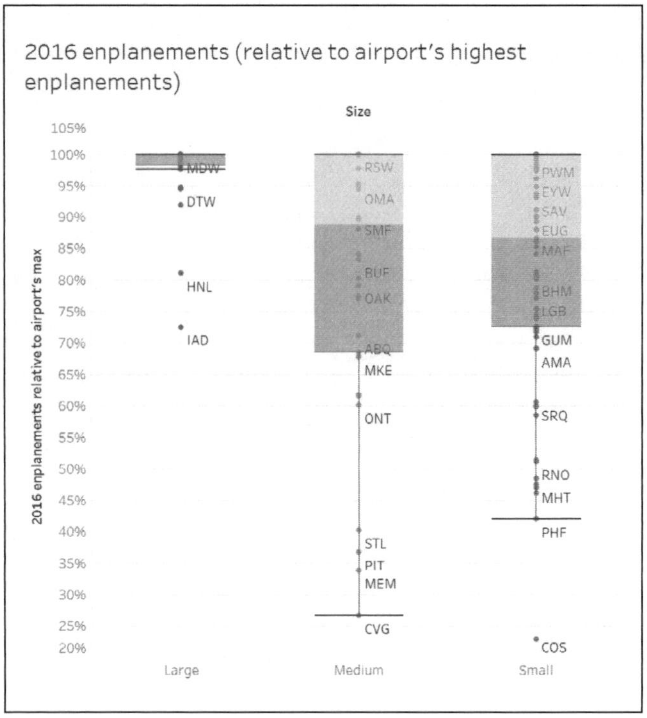

Figure 4. Proxy airport capacity utilization levels by size (normalized 2016 enplanements against historical maximum enplanements).

capacity in terminal facilities. Clearly, even though this metric will tend to underestimate the airport's true excess capacity levels, the resulting CUP volatility distributions could offer valuable information for the robustness of the relative capacity use levels among airports of varying sizes. Persistence in relatively low capacity utilization levels, for example, could identify those airports that could reap higher benefits from the presence of flexibility options in managing their downside risks. Thus, in building activity models, a time series analysis of the activity levels can be especially informative to understand the shock effects over time.

Figure 4 further shows the relative dispersion of enplanement losses among the three size categories and suggests that the conditional value-at-risk losses (average losses following negative shocks) for medium airports appear to be higher. The clustering of losses sustained by Cincinnati/Northern Kentucky (CVG), Memphis (MEM), Pittsburg (PIT), and Saint Louis Lambert (STL) airports in Figure 4, for instance, highlights the concentration of tail risks for this group of airports.

Step 2. Identifying the Nature of Flexibility

The benefits of flexible planning and phasing of airport capital investment projects have been extensively documented (*10*, *16*, *19–23*). In particular, see the overview, references, and cases provided by Chambers (*19*). As such, this step introduces the two previously mentioned flexibility mechanisms that lead to the creation of option values.

Flexibility in Investment Timing. Instead of a now-or-never investment decision implied in static NPV decision rules, a flexible investment approach is adaptive to changes in demand when the expected growth in activity levels does not materialize. In the proposed valuation framework, airport capacity is added incrementally only on observing the quality of the growth trends in demand. The model's decision rule encapsulates the trade-off between the caution in relinquishing the insurance value of the timing option and the risk of missing project revenues because of not having the available capacity when demand arrives.

Flexibility in Design. An airport's ability to adapt to the shifting functional requirements of the aviation industry, such as the constantly evolving security considerations and terminal design guidelines (*24*, *25*), secures the flexibility to accommodate positive jumps in demand if they materialize. The lack of the ability to adjust to new functional requirements, or the absence of easements for future development, for instance, would arguably restrain the fulfillment of renewed interest in the airport in a competitive marketplace. Making the arrival of positive jumps conditional on the presence of flexible design options captures this dynamic, and this feature acts as insurance against losses when negative jumps occur.

As also suggested by the ACRP report on addressing uncertainty in airport decision making, airport operators may have access to various types of mitigation strategies against drops in activity demand levels, including the diversification of air carriers and airports' traffic composition (*12*). When a hub airline leaves an airport, other airlines could fill the excess capacity, and, in some instances, dehubbing may trigger significant shifts in air traffic composition. For example, Amazon's announcement to choose the Cincinnati/Northern Kentucky International Airport (CVG) as an operations hub could be an example for this type of development (*26*).

Step 3. Evaluation of Competing Alternatives

Alternative expansion strategies are evaluated on the basis of their expected NPVs calculated through Monte Carlo simulations. The expansion strategy that yields the maximum NPV is the preferred alternative.

Part II. The Proposed Enplanements Model

Airport activity forecasts are undoubtedly multifaceted, involve considerable levels of complexity and include highly sophisticated estimation methods that often compete with one another (*27–29*). In this study, for the tractability of results, the proposed demand generating equation includes only two separate estimation elements: (1) a trend-stationary geometric Brownian motion with drift; and (2) a jump process, for the baseline demand and rare events, respectively. The rare events may include sharp movements in enplanements because of business cycles, and the departure and arrival of hub airlines. The FAA Terminal Area Forecast provides the trend-stationary drift rate (*30*). Rather than attempting to obtain the best fitting model specification that could potentially cause overfitting of the data, the paper places particular emphasis on identifying and characterizing the uncertainties that afflict the passenger enplanement volatility. Because mean reversion, when present, acts as a systemic dampening mechanism that curbs potential excess capacity losses, a random walk process is assumed.

Passenger enplanement, at time t, (P_t), for a given airport is defined by:

$$\text{Number of enplaned passengers}(P_t) = P_{t-1}\exp\left(\mu_p + \sigma_p W_t\right) \quad (2)$$

where μ_p is the drift term in passenger enplanements, and σ_p is the standard deviation of enplanement growth. The stochastic disturbance in enplanement volumes is represented with a Wiener process, $W_t \sim N(0, 1)$.

Jumps in Airport Activity Levels

This section specifies the probability and magnitude of jumps in activity volumes and the concentration risk that may compound negative shocks.

Negative Jumps. Dehubbing events lead to considerably higher negative jumps than the losses observed in origin and destination traffic volumes. Thus, although the arrival of the negative jumps is modeled to follow a Poisson process, the magnitude of jumps is calibrated to account for the concentration risks of the airports.

Positive Jumps. Positive jumps are only permitted in the presence of flexible design options. This contingent nature of the positive jumps establishes the linkage between the flexibility options in engineering design and airport activity levels. Positive jumps are also expected to occur following a Poisson process. Unlike negative jumps, however, positive jumps are capacity constrained; that is, if there is no available capacity when a positive jump arrives, the additional capacity request is fulfilled to the extent that there is available capacity.

If flexible design options are present, Equation 3 provides the cumulative density function of such events. When no flexibility options for airport design are available, the probability of positive jumps occurring is zero (CDF of positive jumps $F(t;\lambda) = 0$):

$$\text{CDF of positive jumps } F(t;\lambda_{PJ}) = [1 - \exp(-\lambda_{PJ}t)] \quad (3)$$

where λ_{PJ} is the rate parameter that captures the arrival frequency of positive jumps.

Equation 4 expresses the rate magnitude of such positive jumps as:

$$\text{Positive jumps given jump occurs}(\text{PJ}_t) = P_{t-1}[\exp(\mu_{PJ} + \sigma_{PJ} W_t) - 1] \quad (4)$$

where μ_{PJ} is the mean surge in enplanements, and σ_{PJ} is the standard deviation of such jumps in demand.

In contrast, negative jumps are equally likely under both with and without flexible engineering design conditions, and the magnitude of such drops is given by:

$$\text{CDF of negative jumps } F(t;\lambda_{NJ}) = [1 - \exp(-\lambda_{NJ}t)] \quad (5)$$

where λ_{NJ} is the rate parameter that captures the arrival frequency of negative jumps:

$$\text{Negative jumps given jump occurs}(\text{NJ}_t) = P_{t-1}[\exp(\mu_{NJ} + s\sigma_{NJ} W_t) - 1] \quad (6)$$

where s is the scale parameter for concentration risk of connecting passenger enplanements, μ_{NJ} is the mean negative jump in enplanements, and σ_{NJ} is the standard deviation of such drops in demand.

Having defined the three components of the enplanement demand equation, Equation 7 defines the unconstrained enplanement demand. Serviceable demand levels, which are subject to capacity constraints, can be expressed as in Equation 8:

$$\text{Unconstrained enplanement demand}(D_t) = P_t + \text{PJ}_t + \text{NJ}_t \quad (7)$$

$$\text{Fulfilled enplanement demand}(\text{FD}_t) = \begin{cases} C_t & \text{if } (D_t) \geqslant C_t \\ D_t & \text{if } (D_t) < C_t \end{cases} \quad (8)$$

Concentration Risk

Of the many possible ways to simulate jumps in activity, one possible approach is to model origin and destination and connecting passenger demand separately (Equation 9) and assume, for example, that the origin and destination demand tends to be more stable relative to connecting passenger enplanements. The following two variables determine loss magnitude and an airport's exposure to downside risks:

1. Origin and destination (O&D) versus connecting traffic composition
2. Airline concentration in connecting passenger enplanements:

$$\text{Number of enplanements } (P_t) = \text{Origin \& destination enplanements } (\text{OD}_t) + \text{Connecting enplanements } (T_t) \quad (9)$$

The scale parameter, s, for negative jumps can be thought as a concentration index for negative jumps, which is identical to the Herfindahl–Hirschman Index (HHI) in its construction (Equation 10). The HHI is a widely used metric in measuring the competitiveness of an industry by summing the squares of each firm's market share (see the references to the HHI in the Horizontal Merger Guidelines of the U.S. Department of Justice [DOJ] [31]). Although higher values indicate monopolistic market concentration, the presence of many competing firms results in values close to zero. Note that the O&D enplanements shown in Equation 9 are excluded from calculation of the scale parameter for airline concentration (Equation 10):

$$\text{Scale parameter for negative jumps}(s) = \sum_{i=1}^{n} \left(\frac{T_{it}}{T_t}\right)^2 \quad (10)$$

Decision Rule for Expansion Projects

A flexible investment rule may respond to any observed momentum in airport activity trends. This research tests two alternative decision rules to highlight an important trade-off commonly found in expansion projects. Although more conservative investing strategies (waiting for stronger indicators of momentum in demand, for instance) would limit potential losses subsequent to the expansion decision, they might lead to unjustifiably restrictive airport capacity levels. As such, expansion decisions in the model are made based on the period for which the "quality" of demand trends is established, that is, conditional on whether airport activity demand exceeds critical airport capacity in 1 or 2 years in a row. Further refinements to the model could also specify alternative capacity thresholds to trigger the expansion decision. For instance, other decision rules could involve testing the moving average of demand trends against a critical capacity threshold (for an example, see the discussion provided by Cardin on the effect of alternative decision criteria on expansion outcomes [32]).

Equation 11 defines the incremental project revenues by multiplying demand ($\text{FD}_t - C_0$) by a CPE amount, which is the airport operator's only revenue stream in this model.

$$\text{Incremental expansion revenue}(R_t) = \begin{cases} (\text{FD}_t - C_0)\text{CPE} & \text{if } \sum_{i=0}^{t} X_i \geq 1 \\ 0 & \text{otherwise} \end{cases} \quad (11)$$

where CPE revenues are assumed to be net of variable operating expenses, C_0 is the initial airport capacity, and XC is the cost of expansion.

Equation 12 produces the net cash flows, which then can be used to calculate the project's NPV expressed in Equation 13:

$$\text{Net cash flow attributable to expansion}(\text{XF}_t) = R_t - \text{XC}_t \quad (12)$$

$$\text{NPV of the expansion project} = \sum_{i=0}^{n} \frac{\text{XF}_i}{(1+r)^i} \quad (13)$$

where r = the discount factor given for the investment, XF = net cash flow attributable to expansion and XC = cost of expansion.

Finally, Equation 14 calculates the option value in investment timing by differencing the expected NPVs of the flexible and outright expansion strategies (note that both strategies include the second type of option value for the flexibility in design, and thus no adjustment is necessary to remove its value):

$$\text{Option value for flexibility in investment timing} = NPV_{\text{flexible}} - \text{NPV}_{\text{outright}} \quad (14)$$

Because positive jumps are contingent on the presence of the flexibility option in engineering design, Equation 15 provides the expected value of this second type of option value by summing the present value of the revenues attributable to the positive jumps in activity levels:

$$\text{Option value for flexibility in design} = \sum_{i=0}^{n} \frac{\text{PJ}_i\text{CPE}}{(1+r)^i} \quad (15)$$

Part III. Monte Carlo Simulation Example

Model Inputs

Table 3 provides the model inputs that were used to illustrate a hypothetical expansion project by using the enplanements demand equation introduced in Part II.

Model Outputs

This section presents the key outputs of the simulation model. In addition to the model parameters presented above, expansion projects are assumed to take 1 year to complete. In this example, starting from an initial capacity of 1.1 million passengers, on average, the expansion

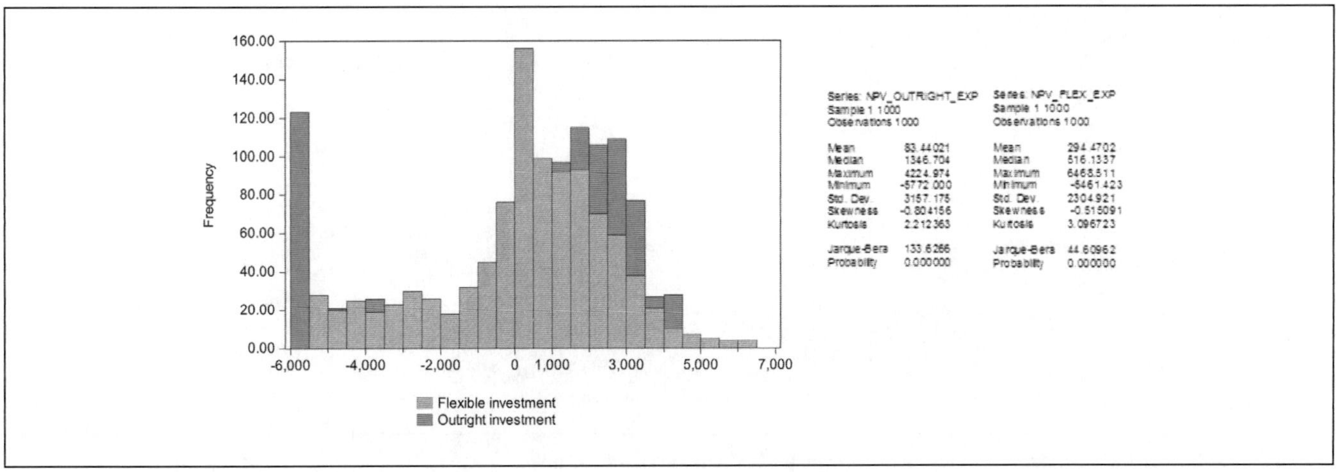

Figure 5. Comparison NPV distributions for flexible and outright investment strategies.

Table 3. Model Inputs

Input type	Input
Existing enplanements capacity (thousands)	1,100
Incremental expansion size (thousands)	200
Default expansion size (thousands)	1,300
Revenue per enplanement (CPE) ($) (net of variable operating costs)	10
Demand variables	
Enplanements trend mean	0.02
Enplanements trend standard deviation	0.01
Positive jump mean	0.10
Positive jump standard deviation	0.05
Negative jump mean	0.20
Negative jump standard deviation	0.10
Expansion costs	
Initial expansion cost ($ thousands)	5,000
Expansion fixed costs/year ($ thousands)	100
Discount rate (%)	5

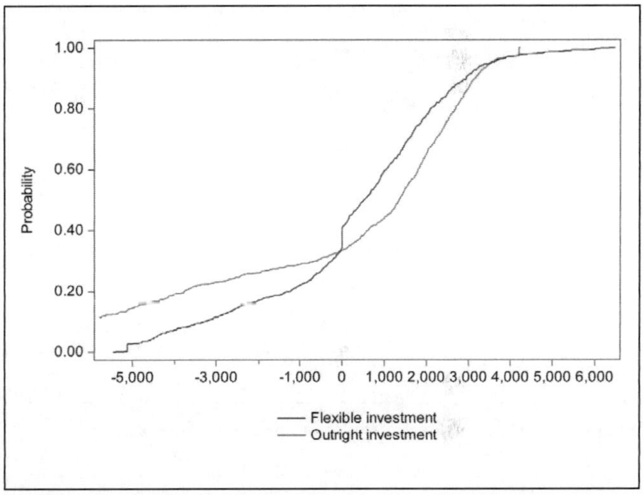

Figure 6. Cumulative density functions of project valuations under alternative expansion strategies.

decision rule generated airport capacity outcomes ranging from 1.1 million to 1.6 million. Based on each of the simulated capacity paths over the analysis period, the NPV distributions for flexible and static expansion strategies were generated by adding the appropriate discounted net cash flows, as expressed in Equation 13.

The comparison of the two alternative expansion approaches (Figure 5) illustrates how the flexible investment strategy can be expected to improve an outright investment strategy. First, note that the expected NPV for the flexible expansion ($294,470) is higher than the outright expansion approach ($83,440). Second, the presence of the option in investment timing significantly reduces downside risks by all but eliminating the clustering of the losses in the far left tail of the outright expansion strategy shown in Figure 5. Further, a flexible approach does extend the right tail of the distribution, which captures the windfall revenues from sustained patterns in enplanement growth paths.

A standard method to compare how different investment alternatives perform against one another is to examine their cumulative distribution functions (CDF) as shown in Figure 6. In general, curves to the right represent improvements over the distributions to the left. Note that, even so, there may not be a Pareto optimal investment strategy between all the alternative strategies. Choosing more conservative expansion strategies may keep potential losses in check, but may prove too cautious in capturing incremental revenues. The trade-off is, thus, between the expected benefits from capping downside losses, the flexibility in adding incremental capacity as growth continues, and the potential losses in revenue while waiting for signs of momentum in activity trends.

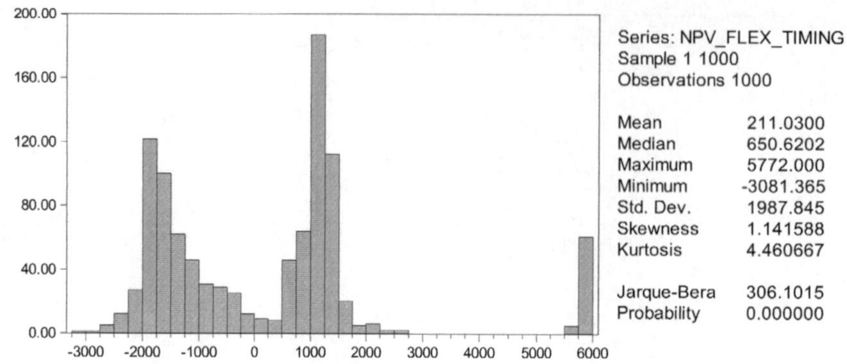

Figure 7. Option value for flexibility in investment timing.

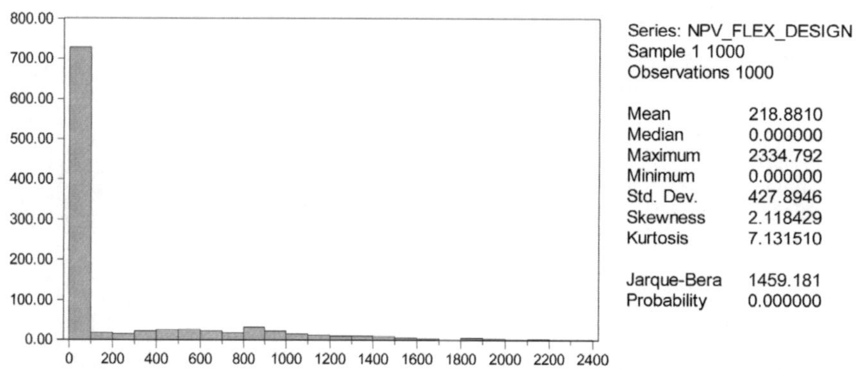

Figure 8. Option value for flexibility in engineering design.

As such, the flexible approach is not without its share of drawbacks. The particular investment rule chosen here proves to be too cautious in expanding capacity because the outright expansion strategy performs better, in part, in capturing enplanement revenues over a significant portion of the positive NPV valuations. As depicted in Figure 6, the flexible expansion strategy offers improvements in the NPV profile by shifting the CDF curve to the right in the negative NPV range but falls short of doing so for the most of the positive range of valuations. The flexible expansion strategy again becomes the preferred alternative, however, in the far positive range of project valuations because of its continued ability to add capacity.

The value of the flexibility option in investment timing then can be obtained by differencing the expected project values under the two strategies shown in Figure 5. The resulting distribution (Figure 7) is the valuation profile for the flexibility option in investment timing. The expected value of this highly asymmetrical distribution provides the option value in flexible investment timing as \$211,030 in this particular example.

The value of the second type of option, flexibility in engineering design, can then be calculated by adding the NPVs of the revenues contributed by positive jumps. Figure 8 shows the NPV profile for this second type of option and results in an option value of \$218,000 (note that without this second type of flexibility option, the project NPV would be negative [\$83,440–\$218,810]).

On balance, therefore, the results of the study demonstrate that airport planners can add significant value from a real options-based valuation framework. Specifically, a flexible approach that monitors and responds to trends in demand is shown to improve the NPV profile of an expansion project. Simulation results also confirm that a flexible expansion approach can provide improvements over outright expansion strategies by simultaneously limiting airports' downside exposures and letting them benefit from sustained growth patterns in activity demand through a series of incremental expansions.

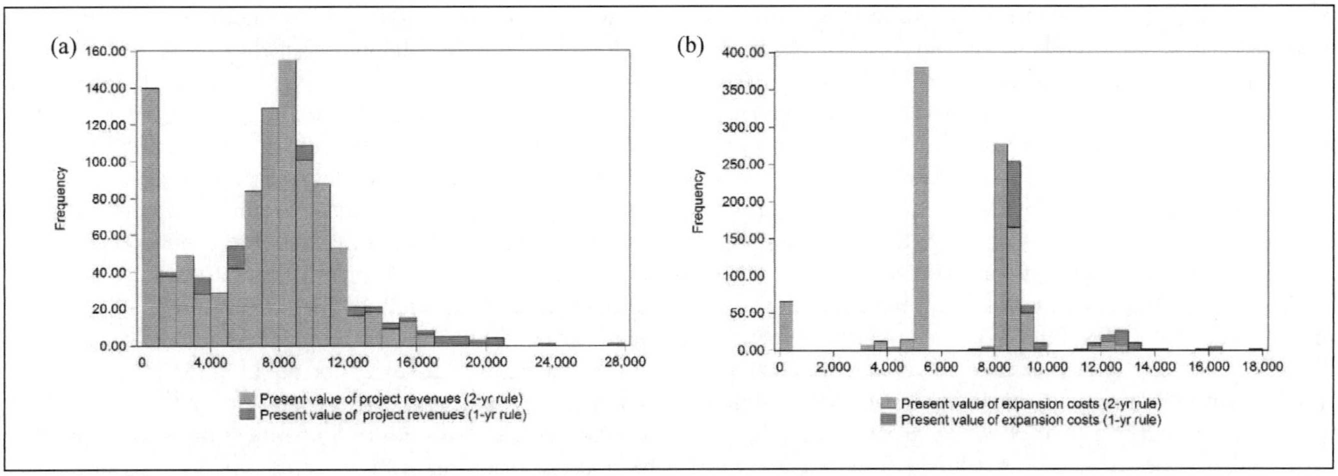

Figure 9. Effects of alternative flexible expansion strategies: (*a*) revenue; (*b*) investment and operating cost distributions.

The elimination of the extreme left-tail losses, especially, corroborates the potential value of the timing options in avoiding what would be unprofitable expansion projects shortly after the arrival of significant drops in demand. Further, as the positive and significant valuation of the second type of option suggests, airports' ability to convert excess capacity into new uses is expected to act as a safety net when activity levels fail to converge to their pre-shock levels and continue to accumulate.

Finally, considering an alternative flexible expansion strategy (replacing the 2-year rule employed thus far), which triggers investment every year capacity falls behind demand, can provide an insight into the effects of using alternative decision rules on the project's expected valuation profiles. Figure 9 shows the expected revenue and capital expenditure distributions resulting from adopting this less conservative expansion approach.

As Figure 9 indicates, the 1-year rule, on average, results in both a higher number of expansion decisions and higher revenues because of the relaxed capacity constraints. However, increased revenues from this relatively aggressive expansion strategy fall short of offsetting the additional expansion costs incurred. A comparison of the cumulative density functions of the two NPV profiles confirms that the 2-year rule is indeed superior to the alternative 1-year rule (Figure 10).

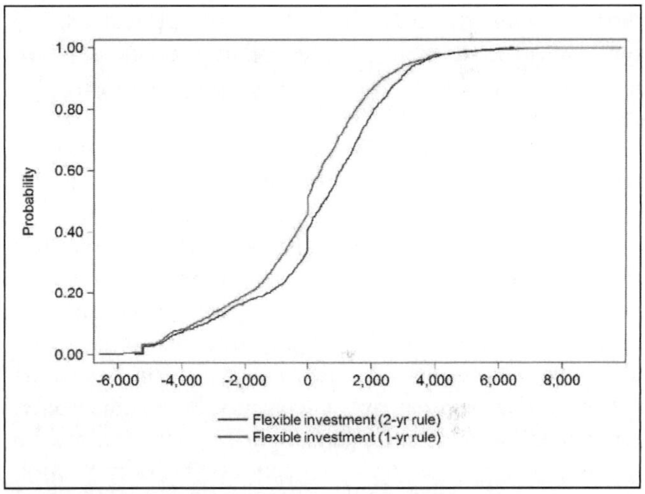

Figure 10. Comparison of NPV profiles under alternative flexible expansion strategies.

Conclusion

Given the sizable uncertainties in future demand forecasts (*17*), and the material opportunity costs associated with overinvestment, flexibility options in planning airport capital improvement projects become paramount. In fact, such flexibilities may create economic value that the traditional NPV analyses fail to capture (*33*). Further, when capital expenditures are irreversible, and excess capacity levels are indicators of inefficient allocation of capital resources, the simulation analysis conducted here identifies two mechanisms that can create value for an airport's capital improvement program. First, a flexible investment rule that monitors and responds to what are believed to be sustained trends in airport activity volumes adds value both by limiting downside losses and by furnishing the flexibility for continued capacity increases. Second, the availability of flexible design options curtails potential losses by backfilling existing excess capacity levels with new business activity.

Although the benefits of such flexible approaches in airport planning are well known, this study shows that option valuations should pay closer attention to the volatility of activity levels both on a yearly and a long-term basis. The discussion demonstrates, for instance, that valuation models can employ a scale parameter to measure an airport's concentration risk in its connecting enplanement activity levels to accomplish this goal. In this setting, the paper expects the concentration risk to be at its highest when only a few airlines make up an airport's connecting traffic business, and this volume represents a substantial share of overall airport activity. Unlike the O&D enplanements that are assumed to be relatively stable, losses in connecting traffic are assumed to exhibit no mean reversion properties.

The results of the simulation example also show not only that the two flexibility mechanisms are capable of increasing project values relative to an outright expansion strategy, but that without the presence of flexibility options the expansion project is not feasible, that is, the project NPV is negative. These results add to the body of evidence on the critical nature of flexibility options in design and planning of airports. Therefore, by mitigating airports' exposures to downside risks, flexible investment approaches may also be of use in justifying increases in total airport investment volumes because of the creation of economic value that would not be present under rigid design strategies.

Like all models, the proposed valuation model can be useful to the extent that simplification of reality is balanced with the reasonableness of its assumptions and their inherent logic. As such, the incremental expansion strategy tested here may not always be the preferred method of adding capacity. The advantages of outright expansion approaches (e.g., savings in construction costs because of economies of scale and disruption to airport operations) may often overwhelm any expected benefits from flexibility options. In addition, uncertainty in future financing costs may favor immediate construction of facilities. Moreover, the simulation exercise also illustrates that whereas the option in expansion timing diminishes downside losses and increases the upside revenue potential, options may also introduce costly revenue losses because of the delay in making expanded facilities operational. The net result of these two countervailing dynamics depends on the chosen investment rule to trigger expansion decisions. For instance, although more conservative investment rules than the one considered here would further eliminate downside losses, they could also lead to even higher lost revenues.

The paper further provides a finding that relates airport sizes to potential option values. Based on an initial analysis of the enplanement growth rates for the 140 largest U.S. airports, the paper argues that not only have large airports outperformed medium and small airports in average growth rates, but these airports experienced lower variance of enplanement growth rates. In contrast to the capacity use trajectories at large airports, the results also suggest that medium size airports may suffer from higher and more persistent downside risks. Despite showing relatively stable enplanement levels compared with small airports, therefore, medium airports seem to be exposed to considerably higher losses as they tend to accumulate negative shocks to their activity levels, without the mean-reverting behavior that the two other types of airports seem to follow. Therefore, the observed higher conditional value-at-risk values are in agreement with the random walk assumption made in the proposed enplanement growth model. A promising future research area and a potential explanation for this finding is likely to be related to the nature and persistence of connecting passenger traffic, which varies across different groups of airport size. That is, whereas large airports are expected to quickly recover from losses in connecting traffic volumes because of their gateway status, smaller airports lack either the ability to make up for their losses or do not have any significant connecting traffic volumes in the first place. Finally, this nonlinear relationship between airport size and average sustained losses is expected to provide valuable insights for planners and operators in forecasting the operating efficiency of expanded airport facilities.

Acknowledgments

This research was supported by the 2016–2017 Graduate Research Award Program on Public-Sector Aviation Issues. The author thanks sincerely Dr. Douglas Gransberg for his continued support and guidance throughout the course of the research. The author also gratefully acknowledges the valuable advice and feedback received from the program mentors, Dr. Richard de Neufville, Eric Amel, Randall Berg, Dr. David Byers, and Thomas Cuddy, and gives thanks for the assistance received from the ACRP staff, and the program manager, Lawrence Goldstein.

Author Contributions

The author confirms contributions to the paper as follows: study conception and design: IK; data collection: IK; analysis and interpretation of results: IK; manuscript preparation: IK. The author reviewed the results and approved the final version of the manuscript.

References

1. Schwartz, E. S., and L. Trigeorgis (eds.). *Real Options and Investment under Uncertainty: Classical Readings and Recent Contributions*. The MIT Press, Cambridge, MA, 2004.
2. Trigeorgis, L. *Real Options: Managerial Flexibility and Strategy in Resource Allocation*. The MIT Press, Cambridge, MA, 1996.
3. Amram, M., and N. Kulatilaka. *Real Options: Managing Strategic Investment in an Uncertain World*. Oxford University Press, Oxford, UK, 1998.
4. Copeland, T., and V. Antikarov. *Real Options: A Practitioner's Guide*. Texere, Knutsford, UK, 2003.

5. Shockley, R. L. *An Applied Course in Real Options Valuation.* Thomson Higher Education, Boston, MA, 2007.
6. de Neufville, R., and S. Scholtes. *Flexibility in Engineering Design*, The MIT Press. Cambridge, MA, 2011.
7. de Neufville, R. Low-Cost Airports for Low-Cost Airlines: Flexible Design to Manage the Risks. *Transportation Planning and Technology*, Vol. 31, No. 1, 2008, pp. 35–68.
8. Hengels, A. *Creating a Practical Model using Real Options to Evaluate Large-Scale Real Estate Development Projects.* Doctoral dissertation. Massachusetts Institute of Technology, Cambridge, MA, 2005.
9. de Neufville, R. Management of Multi-Airport Systems: A Development Strategy. *Journal of Air Transport Management*, Vol. 2, No. 2, 1995, pp. 99–110.
10. Horonjeff, R., F. McKelvey, W. Sproule, and S. Young. *Planning and Design of Airports*, 5th ed. McGraw-Hill, New York, NY, 2010.
11. de Neufville, R., A. Odoni, P. Belobaba, and T. Reynolds. *Airport Systems, Planning, Design, and Management*, 2nd ed. McGraw-Hill, New York, NY, 2013.
12. Kincaid, I., M. Tretheway, S. Gros, and D. Lewis. *ACRP Report 76: Addressing Uncertainty about Future Airport Activity Levels in Airport Decision Making.* Transportation Research Board of the National Academies, Washington, D.C., 2012.
13. Federal Aviation Administration. *Airport Master Plans: AC 150/5070-6B.* FAA Advisory Circular, 2005. https://www.faa.gov/documentLibrary/media/Advisory_Circular/150-5070-6B-Change-2-Consolidated.pdf. Accessed June 30, 2017.
14. Flyvbjerg, B., M. K. Skamris Holm, and S. L. Buhl. How (In)accurate Are Demand Forecasts in Public Works Projects? The Case of Transportation. *Journal of the American Planning Association*, Vol. 71. No. 2, 2005, pp. 131–146.
15. de Neufville, R., S. Scholtes, and T. Wang. Real Options by Spreadsheet: Parking Garage Case Example. *Journal of Infrastructure Systems*, Vol. 12, No. 2, 2006, pp. 107–111.
16. Caves, R. E., and R. Gosling. *Strategic Airport Planning.* Pergamon, Oxford, UK, 1999.
17. FAA. *Terminal Area Forecast.* https://www.faa.gov/data_research/aviation/taf/media/taf_summary_fy_2016-2045.pdf. Accessed June 30, 2017.
18. Mun, J. *Modeling Risk: Applying Monte Carlo Simulation, Real Options Analysis, Forecasting, and Optimization Techniques.* John Wiley & Sons, Hoboken, NJ, 2006.
19. Chambers, R. D. *Tackling Uncertainty in Airport Design: A Real Options Approach.* Master of Science thesis. MIT Technology and Policy Program, Cambridge, MA, 2007.
20. *Capacity Needs in the National Airspace System.* Federal Aviation Administration, Washington, D.C., 2004. https://www.faa.gov/airports/resources/publications/reports/media/NAS_needs.pdf. Accessed on June 30, 2017.
21. de Neufville, R. Designing Airport Passenger Buildings for the 21st Century. *Transport Journal, UK Institution of Civil Engineers*, Vol. 111, 1995, pp. 83–96.
22. de Neufville, R. *Amsterdam Multi-Airport System Policy Guidelines.* Report Prepared for Amsterdam Airport, Schiphol, The Netherlands, 1995. http://ardent.mit.edu/airports/ASP_papers/multi-airport%20systems%20policy%20guidelines.PDF2047&rep = rep1&type = pdf. Accessed June 30, 2017.
23. Burghouwt, G. *Airline Network Development in Europe and Its Implications for Airport Planning.* Ashgate Publishing Company, Burlington, VT, 2007.
24. Odoni, A., and R. de Neufville. Passenger Terminal Design. *Transportation Research Part A: Policy and Practice*, Vol. 26, No. 1, 1992, pp. 27–35.
25. de Neufville, R., A. G. de Barros, and S. C. Belin. Optimal Configuration of Airport Passenger Buildings for Travelers. *Journal of Transportation Engineering*, Vol. 128, No. 3, 2002, pp. 211–217.
26. CVG. *Airport Press Release.* http://www.cvgairport.com/about/news/2017/01/31/amazon-to-create-2-000-jobs-at-new-cvg-hub. Accessed June 30, 2017.
27. Bhadra, D., and R. Schaufele. Probabilistic Forecasts for Aviation Traffic at FAA's Commercial Terminals: Suggested Methodology and Example. *Transportation Research Record: Journal of the Transportation Research Board*, 2007. 2007: 37–46.
28. Spitz, W., and R. Golaszewski. *ACRP Synthesis 2: Airport Aviation Activity Forecasting.* Transportation Research Board of the National Academies, Washington, D.C., 2007.
29. *Transportation Research Circular E-C040: Aviation Demand Forecasting: A Survey of Methodologies.* TRB, National Research Council, Washington, D.C., 2002. http://onlinepubs.trb.org/onlinepubs/circulars/ec040.pdf.
30. Federal Aviation Administration. *Aerospace Forecasts, Fiscal Years 2016–2045.* Federal Aviation Administration, Washington, D.C., 2016. https://www.faa.gov/data_research/aviation/aerospace_forecasts/. Accessed on June 30, 2017.
31. U.S. Department of Justice. *Horizontal Merger Guidelines.* https://www.justice.gov/atr/herfindahl-hirschman-index. Accessed on June 30, 2017.
32. Cardin, M. A. *Facing Reality: Design and Management of Flexible Engineering Systems.* Master's thesis. Massachusetts Institute of Technology, Cambridge, MA, 2007.
33. Pindyck, R. Irreversible Investment, Capacity Choice, and the Value of the Firm. *The American Economic Review*, Vol. 78, No. 5, 1998, pp. 969–985.

The ACRP Selection Panel for the Graduate Research Award Program on Public-Sector Aviation Issues peer-reviewed this paper (18-04039).

Article

TRR
JOURNAL OF THE TRANSPORTATION RESEARCH BOARD

Transportation Research Record
2018, Vol. 2672(29) 82–92
© National Academy of Sciences:
Transportation Research Board 2018
Article reuse guidelines:
sagepub.com/journals-permissions
DOI: 10.1177/0361198118799705
journals.sagepub.com/home/trr
SAGE

Airfield Pavement Damage Evaluation Due to New-Generation Aircraft Wheel Loading and Wander Patterns

Priyanka Sarker and Erol Tutumluer[1]

Abstract

This paper presents a stress-history-based approach to predict the deformation basins of airport pavements subjected to heavy aircraft loading applied in sequential wanders. Multi-depth deflectometer data from full-scale aircraft landing gear tests conducted at the National Airport Pavement Test Facility built by the Federal Aviation Administration are used to create individual pass residual deformation transverse profiles. The computed residual deformation profiles are further corrected for stress-history effects to predict rut in the selected test sections. The developed model focuses on using the previous load location and stress history of the soil element to develop the deformations in that element. Despite the unavailability of the surface transverse profile data measured in the field at different passes, the initial attempt of the model can closely predict the deformation profile similar to width and shape expected in the field. And after the stress-history effects are accounted for, the initially calculated rut depth decreases significantly to match the final contour basin of the test sections extracted from the post traffic trenching. The advantage of using the stress-history-effects-based rut prediction tool is that it can allow any combination of wander positions and sequences of load applications to be accounted for their effects on the final surface rut development.

Higher load levels and load wanders (later movement of aircraft) due to non-channelized air traffic loading is a phenomenon primarily associated with airfield pavements because unlike highway pavements, wander widths in airport pavements contain a zone with 75% of the aircraft centerlines (1.15 standard deviations on either side of the mean value) (*1*). Though wander reduces the number of repetitions of maximum load applied to the most heavily trafficked pavement location, several studies have indicated that wander does not necessarily increase the pavement life. Rather it causes constant particle shuffling and rearrangements in unbound layers, ultimately leading to a weaker and more unstable granular layer (*2–6*). In addition, more frequent use of new-generation aircraft (NGA) combined with rapid airline traffic growth has made the airport pavements more exposed to various types of heavy aircrafts with complex multiple wheel gear landings coupled with wander. So, a thorough understanding of pavement damage mechanisms considering all the traffic circumstances, i.e., a diverse mix of airplanes with complex landing gears coupled with wander, has become necessary, to design, build, and maintain sustainable airfield pavement sections.

Currently, the failure models used in the FAA's flexible pavement thickness design procedures consider maximum vertical strain at the top of the subgrade (for rutting) and the maximum horizontal strain at the bottom of all asphalt layers (for cracking) as the failure criteria (AC 150/5320-6E) (*7*). Though subgrade rutting is taken as the controller mode of structural shear failure, full-scale testing data has shown that most of the rutting in the pavement system is typically contributed by unbound granular layers whereas the rutting in subgrade is found to be minimal (*8–10*). This behavior of the unbound layers can be attributed to wanders because National Airport Pavement Test Facility (NAPTF) full-scale testing data has also indicated the detrimental effects of wanders on the performance of unbound granular (*8, 11, 12*). It was seen that some of the residual deformation caused by a single pass was recovered due to subsequent load applications offset by wander during a complete trafficking wander that included a sequential wander pattern covering approximately 82% of all traffic from a standard normal distribution curve of real-world

[1]Department of Civil and Environmental Engineering, University of Illinois at Urbana-Champaign, Champaign, IL

Corresponding Author:
Address correspondence to Erol Tutumluer: tutumlue@illinois.edu

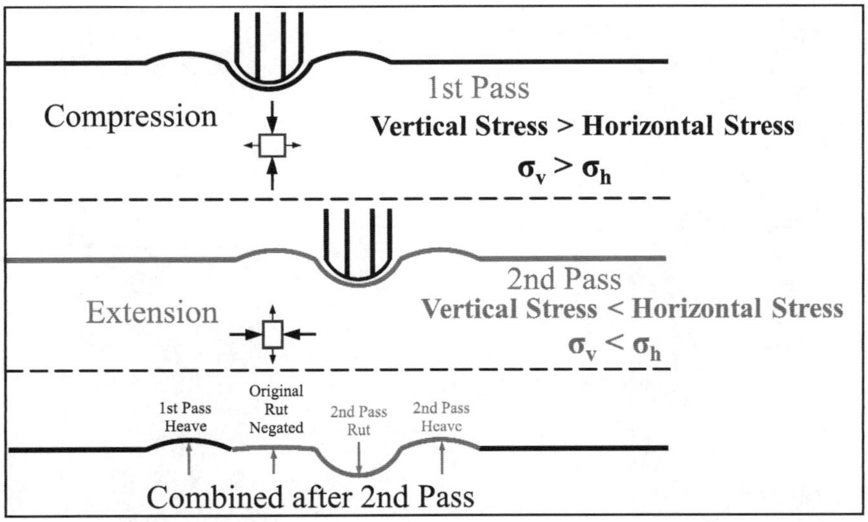

Figure 1. Schematic explaining the rut profile development from an offset wheel.

taxiway traffic (*11, 12*). Figure 1 provides a simple diagram of the observed behavior and shows how the stress in a soil element offset from a load can change with a moving wheel.

Also, dynamic pavement response studies conducted at the University of Illinois at Urbana-Champaign to investigate the deformation behavior of unbound aggregates due to wanders indicated that the unbound aggregate particles moved because of the constantly changing load application lane (*3–6*). This movement negated the consolidation or shakedown expected in unbound aggregate layers under continuous loadings. These findings were further cemented by the comparison of channelized traffic and traffic with wander, which showed that traffic with wander was more detrimental to the unbound aggregate layers due to the increased movement and rearrangement of particles in the unbound aggregate layers.

Thus, this study aimed to investigate the effect of aircraft wanders on the rut accumulation in the pavements. This paper includes NAPTF full-scale pavement test section findings from the analyses of multi-depth deflectometer (MDD) data and deformation trends observed in the unbound aggregate layers due to applied aircraft gear loading with wander.

Objective and Scope

As previously mentioned, recent NAPTF full-scale testing data has shown that most of the rutting in a flexible pavement section was contributed by the unbound layers whereas rutting in the subgrade material was minimal. And constant particle rearrangements due to wander plays a major role in damaging the unbound layers. Addressing these concerns, the objectives of the paper was to develop a stress-history-based MDD data analyses approach to predict the deformation potentials of airport pavements subjected to heavy aircraft loading with wanders.

Selected Test Sections

The FAA built the NAPTF to generate full-scale testing/trafficking data to support the investigation of the performance of airport pavements subjected to complex gear loading configurations of NGA loading applied with wheel wander. The NAPTF consists of a 900 ft (274 m) long and 60 ft (18.3 m) wide test track along with a rail-based test vehicle built with two carriages. Each carriage in the vehicle can be equipped with two load axles with each axle capable of carrying 20 wheels. Maximum total load capacity of the testing vehicle is 1.3 million pounds (5.78 MN) that can be applied with various combinations of landing gear and wander sequences. Full-scale testing conducted at the NAPTF is denoted by construction cycles, and one of the most recent series of full-scale flexible pavement tests conducted with available field data is referred to as the Construction Cycle 5 (CC5) test, part of which is the subject of the results presented in this paper.

In this study, two test sections were selected from the NPATF CC5 test. These pavement sections were constructed with two different subbase materials which consisted of dense graded aggregate (DGA), denoted as Section 1, and crushed quarry screenings (P154 materials), denoted as Section 2. The coarse aggregates in the DGA mix were much larger than those of the crushed

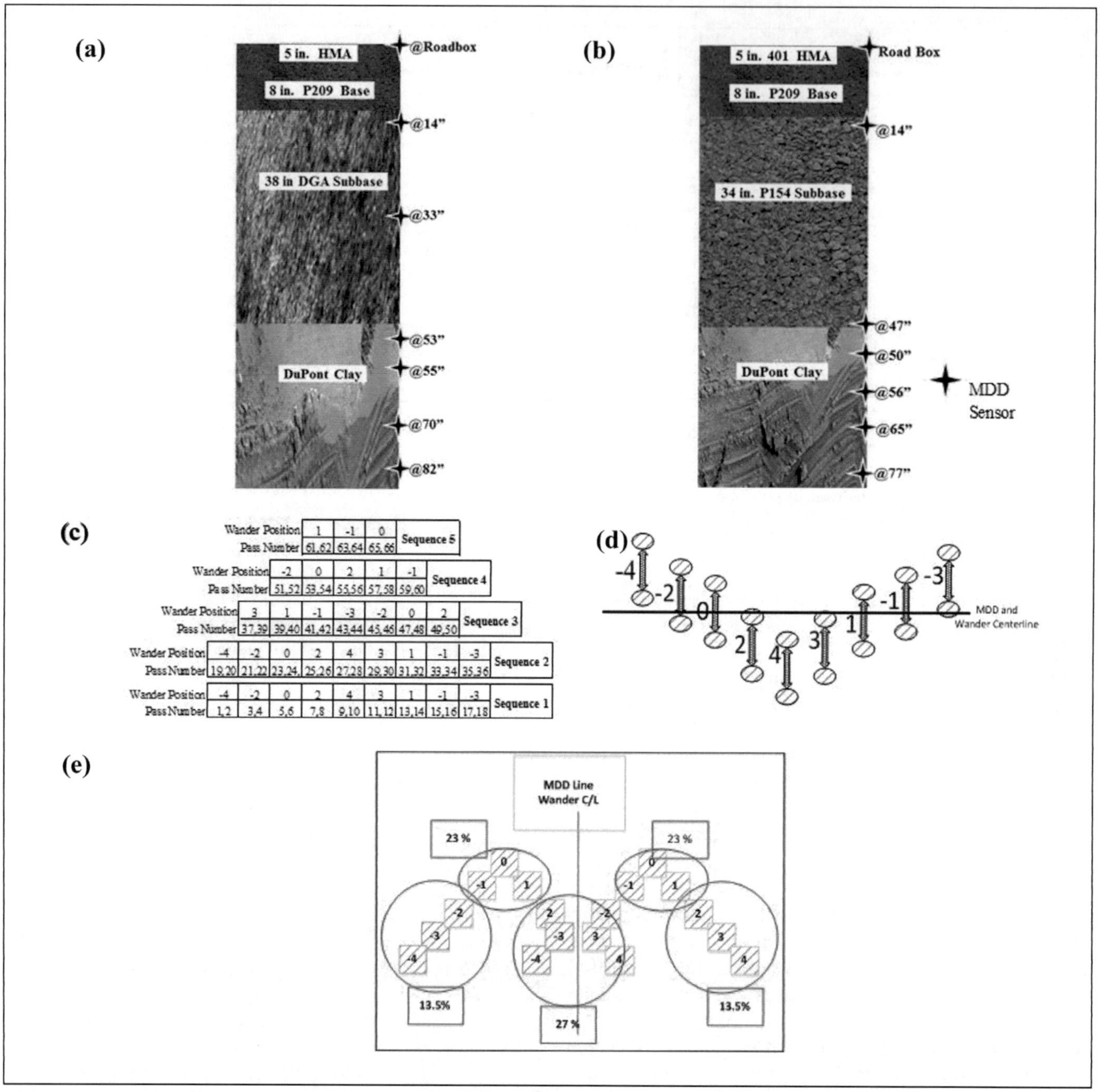

Figure 2. (*a*) Layer configurations of Section 1; (*b*) layer configurations of Section 2; (*c*) applied wander sequence; (*d*) 6-wheel gear centerline position for each wander position; and (e) distribution of the traffic.

quarry screenings mix. The pavements were built on a low California bearing ratio (CBR) DuPont Clay subgrade (average CBR of 3.3 at an average moisture content of 37.6%). An 8 in. (203 mm) depth of P209 crushed stone base was placed on the subbase layers and the sections were completed with the placement of a 5 in (127 mm) thick hot mix asphalt (HMA) surface layer. Figure 2*a* and *b* show the configurations of the selected test sections.

The CC5 test sections were instrumented with sensors such as MDD. An MDD consists of a set of displacement transducers (DT) to measure the movement of a stack of a set of snap ring anchors embedded in the pavement layers at predetermined locations. The surface DT measures the displacement relative to an anchor embedded at a deep location where no significant displacement should occur, and the other DTs basically measure the movement of the snap ring anchors relative to the surface DT.

Therefore, only the surface sensor readings are absolute movements of the pavements, the actual movements of the other DTs need to be calculated by using the movement of the surface DT. The facility was also equipped with a computer-controlled data acquisition system to store the pavement response and performance data collected using MDD sensors as the test vehicle with six wheels of dual-tridem landing gears traveled along the section at a travel speed of 5 mph (8 km/h). The traffic loading was applied in a simulated wander pattern (See Figure 2*c*) that represented the taxiway distribution for design consisting of 66 repetitions, 33 travelling west and 33 travelling east, arranged in nine equally spaced wander positions (or tracks) at intervals of 10.25 in. (260 mm) (See Figure 2*d*). The dual wheel spacing used in 6-wheel landing gear was 54 in. (1372 mm). Figure 2*e* shows the traffic distribution across the pavement sections, and as can be seen, the area consisting of MDD centerline and the wander centerline received most of the traffic (around 27% of the total traffic). Wheel loads were initially set at 59,000 lbs (262.5 kN) with a tire pressure of 243 psi (1,675 kPa) for Section 1. After 12,936 passes the load was increased to 65,000 lbs (289 kN) per wheel and at 18,612 passes it was again increased to 70,000 lbs (311 kN). Section 2 was always trafficked at 70,000 lbs (311 kN). Section 1 was trafficked up to 22,572 passes whereas Section 2 received around 12,936 passes. The next section will discuss the research approach adopted in this study to develop the stress-history-based rut prediction tool by analyzing the CC5 MDD database.

Development of the Stress-History-Based Rut Prediction Tool

Analysis of the NAPTF CC5 MDD Data

As discussed earlier, two sections of varying layer configurations built with two different subbase materials from the CC5 test database were selected in this study. The MDD database had around 12,000 passes of the test vehicle for each section. Each pass consisted around 300 data points recorded for each sensor and there were six sensors per MDD stack. In total, each section had around 21.6 million records. The database had this information divided into individual sensor reading with 300 data records as comma separated values (CSV). So, the first step in the data analyses involved converting the CSV MDD sensor records into individual layer deformation values. And this was achieved by writing a program in Python that could read the complete MDD database, select the sensors by number, extract the selected sensor data, parse the CSV values, convert them into individual layer movements (by subtracting the lower sensor reading from the upper sensor reading), and calculate the residual deformation values (final value of the sensor minus the original value of the senor). The next step in the analyses included cleaning the processed deformation data. Careful attention in this step was given to ensure that all the sensors were showing reasonable displacement values i.e., the top layers were deforming more than the bottom layers. After the cleaning and sorting, Section 1 and Section 2 were left with only around 6,000 and 3,500 passes, respectively.

Once the MDD database was processed, cleaned, sorted, and organized they were further separated by wander position (track −4 to 4), travel direction (east or west), and wander sequence (sequence 1 to 5). The separation of the residual deflection data for wander position 0 (the most heavily trafficked position) by individual wander sequences and traffic direction for the surface and P209 layer and the DGA layer are shown in Figure 3*a* and *b*, respectively.

Similar results are presented at Figure 3*c* and *d* for Section 2. Please note, the interface of HMA surface and P209 base layer had no MDD sensor installed, hence these two layers were analyzed together. The data separation led to an important observation that the first pass on each wander position in the west to east direction typically caused the most response and the return pass along the same wander position showed significantly less residual deflection. This finding clearly indicated the presence of shakedown in the unbound aggregate layers. Figure 4 shows the aggregate shakedown concept as divided into three zones by Werkmeister et al. (*13*). These zones are as follows:

1. Zone A—plastic shakedown, is characterized by a quick decrease in the residual deflections which eventually leads to the layer showing no further residual deformation with additional load repetitions.
2. Zone B—plastic creep initially goes through a decreasing residual deformation rate but as the number of load cycles increase, the residual deformations rate increases subsequently leading to an incremental collapse.
3. Zone C—which is incremental collapse shows a slower reduction in the residual deformations than range A or B and a quick resurgence of the strain rate after a very limited number of load cycles.

According to the shakedown theory, all unbound aggregates experience some shakedown due to traffic. If shakedown did occur in the unbound aggregate layers during full-scale testing, the percent of residual response by layer would decrease for the subbase layers. But, as observed in Figure 5, where the percent of the residual response by each layer of the pavement system of Section

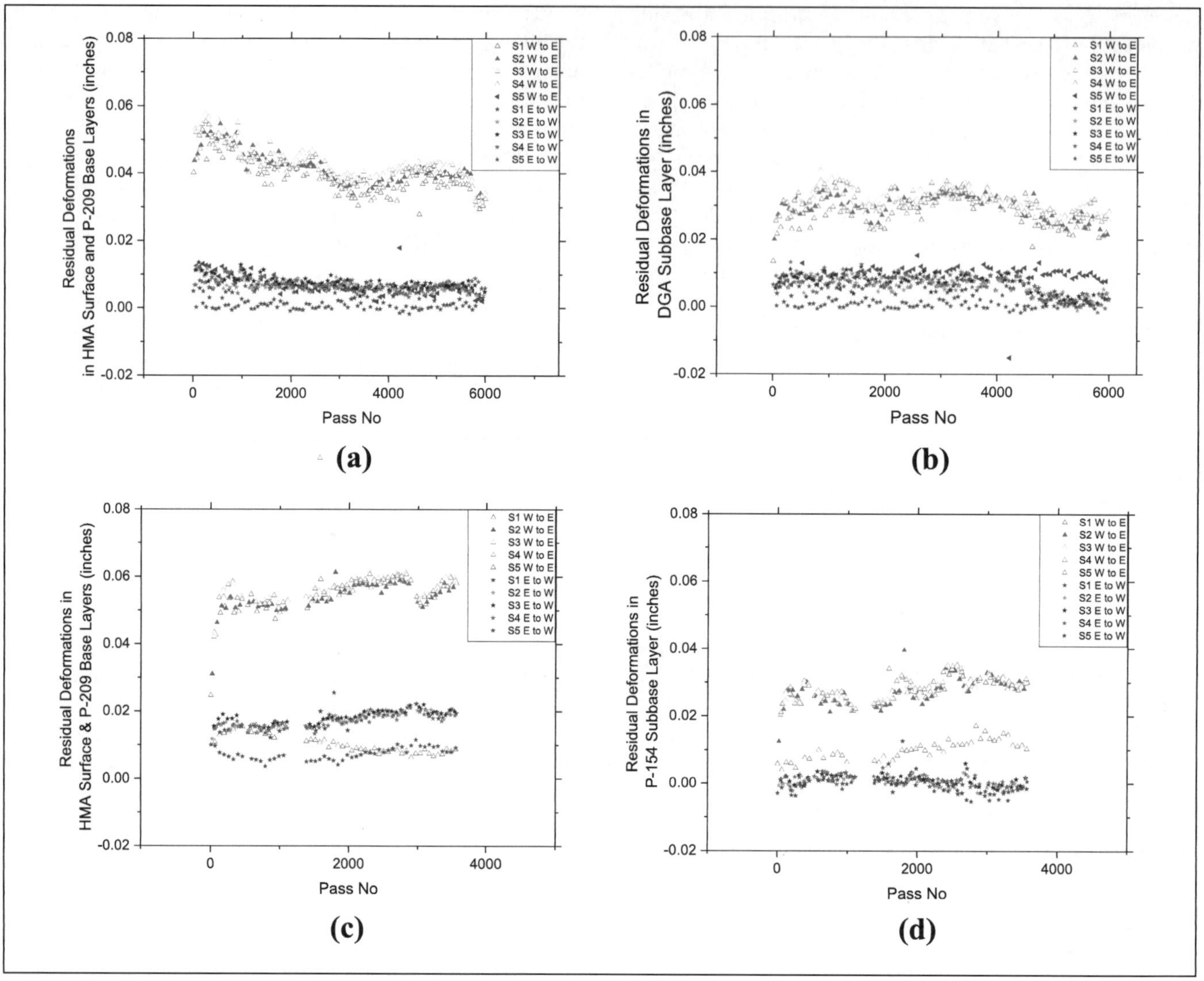

Figure 3. Residual deformation for (*a*) HMA surface and P209 layers of Section 1; (*b*) DGA subbase layer of Section 1; (*c*) HMA surface and P209 layers of Section 2; and (*d*) P154 subbase layer of Section 2 (1 inch=25.4 mm).

1 is shown, this did not happen with increasing pass number. The percent of the residual response remained relatively constant for each layer as the testing progressed, i.e., 56% for surface and P209, 41% for DGA, and 3% for the subgrade which is contradictory to the shakedown theory where the unbound aggregate layers should consolidate and hence decrease their contribution to permanent deformation.

As can be seen in Figure 5, the initial 1,000 passes showed an increase in the percent of residual response for the DGA subbase layers. Interestingly, this increase in the DGA layers was counteracted by a decrease in the percent of residual response by the surface and P209 layers for the same 1,000 passes. This can be attributed to range B behavior where the constant particle rearrangement slowly deteriorated the DGA layer while at the same time the surface and P209 layer consolidated due to traffic. However, overall it can be seen that the percent of residual deflection responses in all the layers were relatively constant. Due to the excessive movements of the layers, the MDD data were only reliable up to 6,000 passes. Similar plot could not be produced for Section 2 due to data unavailability because the MDD data were only reliable up to 3,500 passes.

The influence of wander position and wander sequence on the residual response values can also be further evaluated by combining the data into 66-pass wander pattern. The residual deformation data separated into 66-pass wander pattern are presented in the Figure 6*a* and *b* as individual lines. Figure 6*a* and *b* show the 66-pass wander pattern for the DGA subbase course and surface and P209 layers of Section 1, respectively.

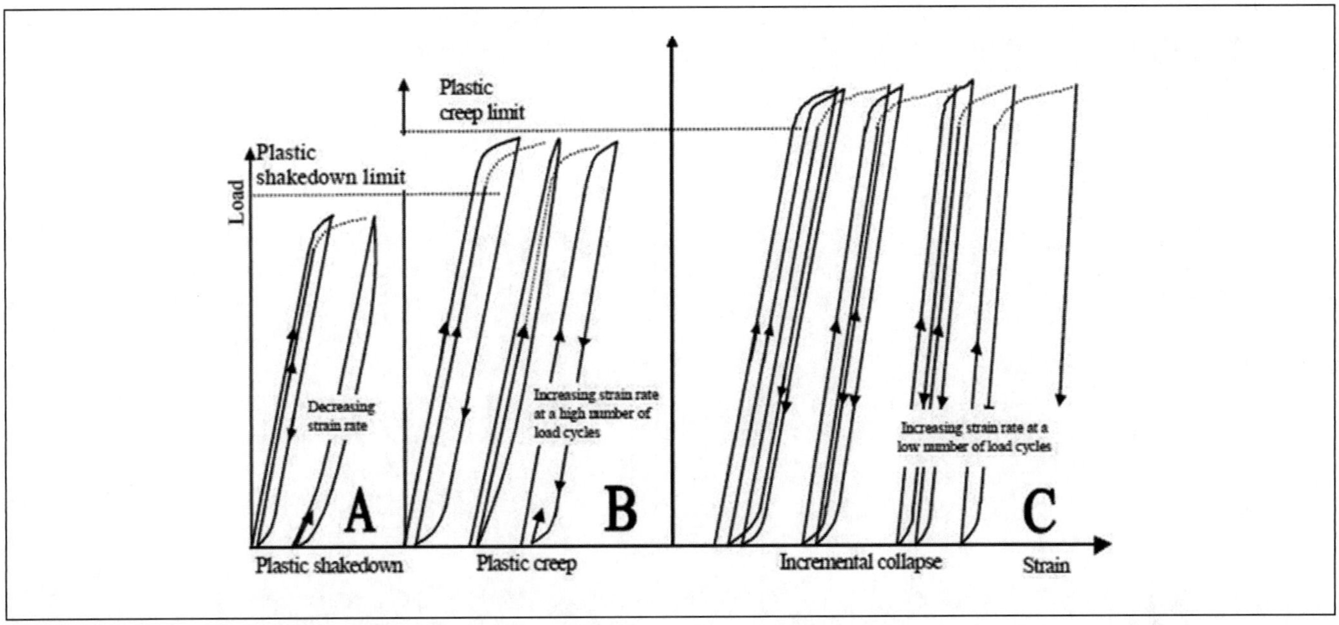

Figure 4. Behavior of granular materials under repeated cyclic load (*13*).

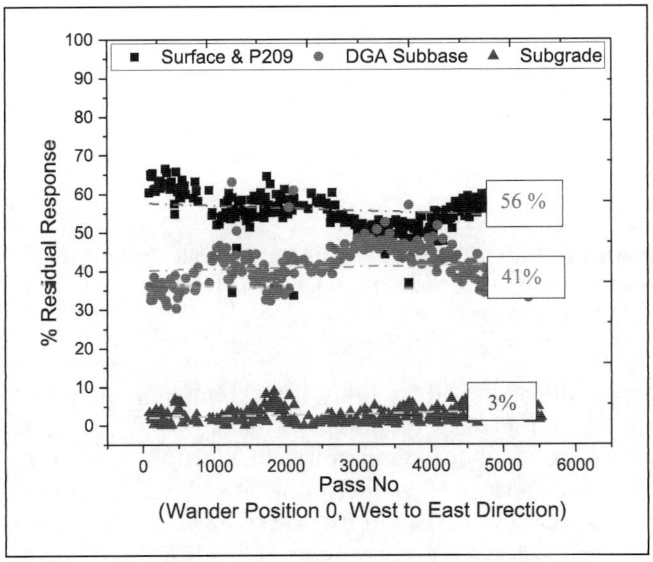

Figure 5. Layer percentage of the total residual response, Section 1, west to east loading direction, wander position 0.

Similar results are shown for Section 2 in Figure 7*a* and *b*. As can be seen from these figures, traffic in the west to east direction on wander position −3 produces the maximum downward residual deformation, which correlates with the position where a wheel is directly over the MDD centerline (See Figure 2*d*). Wander position 2 is the only other wander pattern to produce consistent downward residual deformations and is the second closest wheel load. All the other wander positions contribute to various amounts of rutting or heave.

Wander position 4 is again the offset that causes the most heave of the pavement system and individual layers. Wander position 0 also causes considerable heave to the system. Figures 6*a*–*b* and 7*a*–*b* show that the rut and heave residual deformation values generally increase with the number of passes. Figures 6*a*–*b* and 7*a*–*b* also show that depending on the location of the wander, the pavement sections can either go through rut or heave because the sequential wander pattern is causing constant particle shuffling in the granular layers and not letting shakedown happen. But when the wander pattern is kept narrow, shakedown does happen, which explains the visible reduction in residuals for wander sequence 5. Also, less rutting and less heaving are observed where the traffic is just reversed in the same wander position during the same return pass. The next subsection will discuss using the residual response values to predict the deformations in the selected test sections.

Development of the Deformation Basin

Once it was established that the wander locations and sequence influenced the magnitude of the residual deformations found in the test sections, the next stage in this study was to use this information to create the residual deformation basins. Since the MDDs were not placed transversely across the pavement but rather at a single location on the pavement, the MDD residual response data from different wander positions was used to create a quasi-transverse profile due to a single gear carriage pass. Each of the stationary surface MDD sensor readings measured during trafficking of the nine wander positions

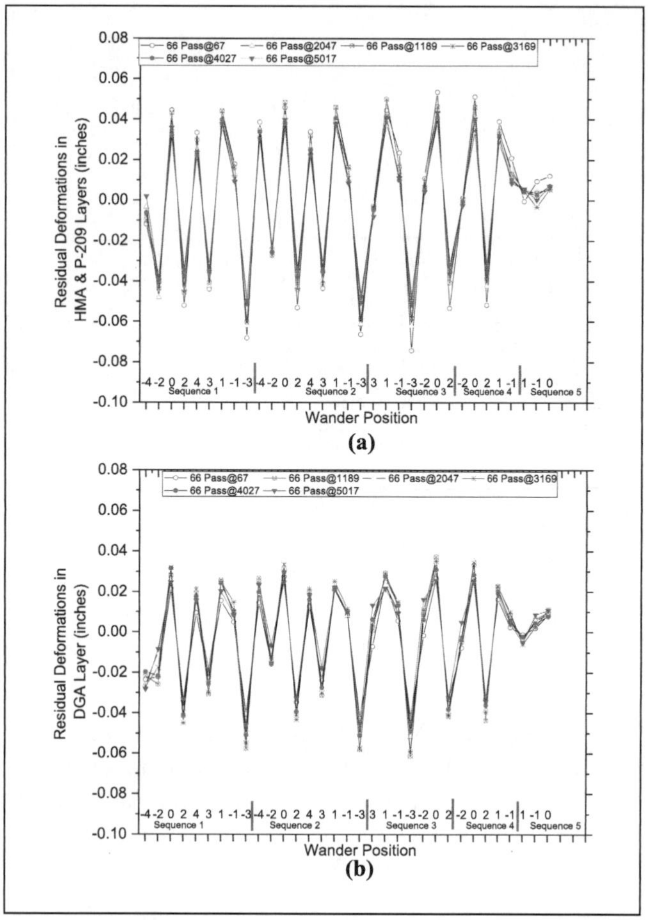

Figure 6. Residual response over 66-pass wander pattern and wander sequence for Section 1: (*a*) HMA and P209 layers; (*b*) DGA subbase layer (1 inch=25.4 mm).

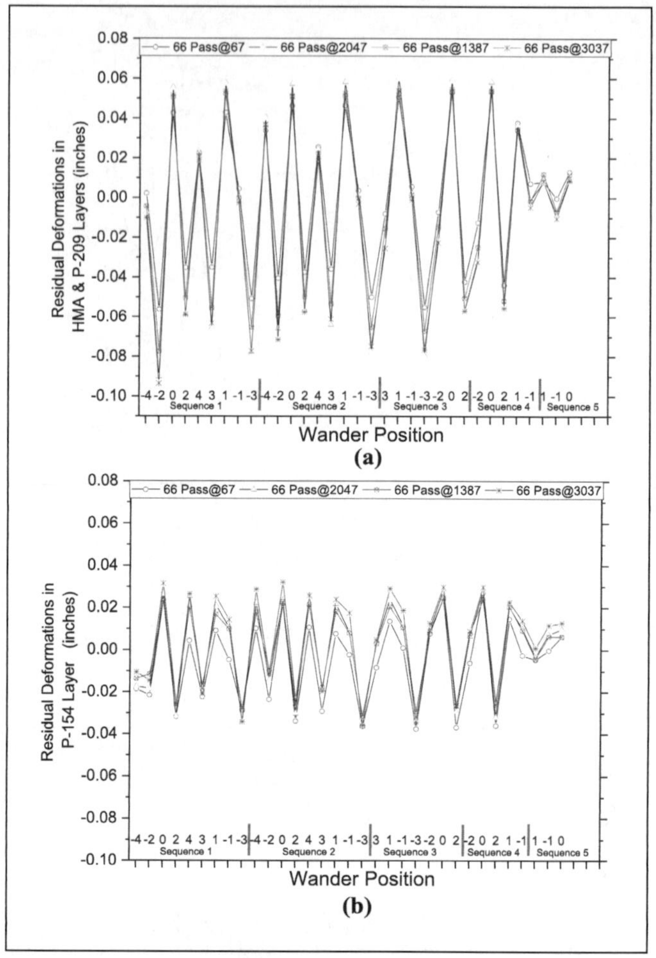

Figure 7. Residual response over 66-pass wander pattern and wander sequence for Section 2: (*a*) HMA and P209 layers; (*b*) P154 subbase layer (1 inch=25.4 mm).

could provide a transverse profile data point. However, since the wander locations were very close together between the gear wheels, a slight difference in the readings could lead to significant changes in the resulting deformation basin, so only the locations that had the maximum downward and upward residual values were used in this study to develop the deformation basin. The methodology of using stationary MDD responses to create deformation basins of pavements has been well documented by Donovan et al. (*14*).

The first step in creating the profile involved determining which MDD readings are critical to developing the transverse profile. As previously described, for both the sections, wander position −3 and wander position 4 readings consistently showed the maximum downward residual and upward deformation values, respectively; except for Section 2 surface layer where wander position −2 had the maximum downward values. However, for the convenience of the analyses, wander position −3 will be considered the critical location with maximum downward residual values. And ander position 4 will be taken as the other critical location. The average of these two values were used at the midpoint of the gear. The point of zero residual deformation was assumed to be at 18 in. (457 mm) outside of wander position 4. Once the critical points were established, it was assumed that if the transverse profile was a mirror image, the critical MDD reading locations could be inverted around the wander centerline and thus, a 118 in. (4,340 mm) wide complete transverse profile was created. Once the width of the transverse profile was known, the values of the residual deformation between the critical points were determined. It was done by assuming that the residual deformation profiles are of parabolic shapes and one critical point was the vertex of a parabola and the other was a point on the parabola. The value of the transverse profile was calculated every 0.25 in. (6.35 mm) and combining the parabolas resulted in a smooth transverse profile created using only the critical points (See Figure 8*a*). Figure 8*a* also shows the critical point locations used to calculate then create the transverse surface profile. As expected, residual

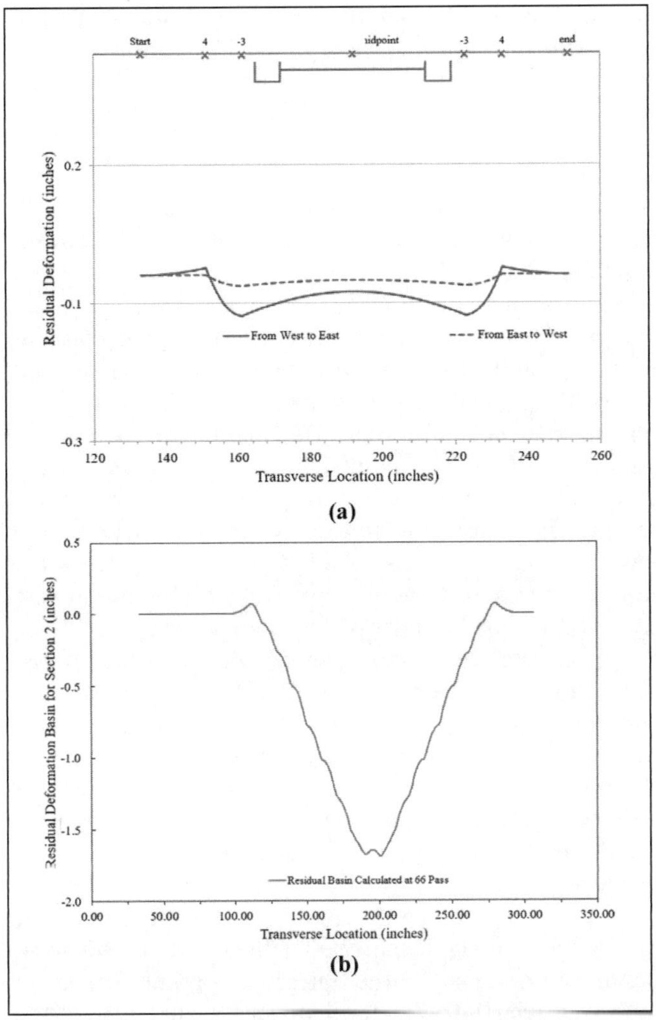

Figure 8. (*a*) Critical points used for transverse profile creation; (*b*) 66-pass residual basin calculated for Section 2 (1 inch=25.4 mm).

deformation recorded during a west to east pass was different than that seen from an east to west pass. Though this method to create the transverse profiles was not completely accurate, it provided valuable insight into the creation of the residual deformation basin. Therefore, it was used in this study to compute the accumulation of the residual deformation basin. Figure 8*b* shows one example of one set of completed transverse profiles for a 66-pass wander pattern calculated by combining the residual deformation - transverse profile from each wander sequence and travel direction. Figure 9*a* and *b* show the residual deformation basin calculated for Section 1 and 2, respectively.

As can be observed from Figure 9*a* and *b*, the summation of the individually determined transverse profiles from each pass of a 66-pass wander pattern led to a bowl-shaped crater basin that matched the contour of the transverse profile as seen from post-trafficking trenching (See Figure 9*c*). It can also be seen that the depth of the transverse profiles created with the MDD readings increased with an increasing number of passes. However, the magnitude was much greater than the actual transverse profile found in the field as shown in Figure 9*c*. That is, if the created transverse profiles were summed over the actual number of complete 66-pass wander patterns covered in 22,572 passes, the rut depth amount was much greater than the measured values (the resulting accumulated rut was almost 80 times the measured amounts). The same trend was observed in the accumulated transverse profiles in the Section 2 as well, where the resulting accumulated rut was almost 30 times the measured amounts. So, though the shapes of the created transverse profiles closely followed the field profiles, when the created transverse profiles were summed for the number of complete 66-pass wander patterns, the magnitudes of the created profiles exceeded those found in the field.

This can be attributed to the fact that amount of deformation in granular layers depends on whether the soil element experienced rut or heave in the last pass; if the element was compressed in the previous pass, it would show more heave in the next offset pass. Likewise, if the element heaved in the last pass, the next pass when the load was directly over the element would make the soil rut more. So, the residual transverse profile from each pass was affected by the residual profile of the previous pass. Therefore, if the stress-history effects were ignored, predictions of transverse surface profile in CC5 tests will be inaccurate. This observation supports the findings documented by Kim and Tutumluer (*2*) about the importance of stress-history effects on the behavior of granular layers and how by including stress-history adjustment factors into the rutting prediction equations, permanent deformations could more accurately be predicted. The next subsection will introduce a stress-history correction approach to predict the rut depths in the selected test sections.

Development of Stress-History Correction Approach

As explained earlier, the final shape of a residual deformation basin using an MDD based transverse profile creation approach can be predicted by knowing the location of the gear wheels for each wander position and number of passes. However, calculating the magnitude of the residual transverse profile across a pavement section based on an individual residual transverse profile is extremely difficult. In a recent study using NAPTF Construction Cycle 1 (CC1) and Construction Cycle 3 (CC3) database, a methodology was developed to predict the final transverse profile from multiple passes by using the residual transverse profile caused by an individual

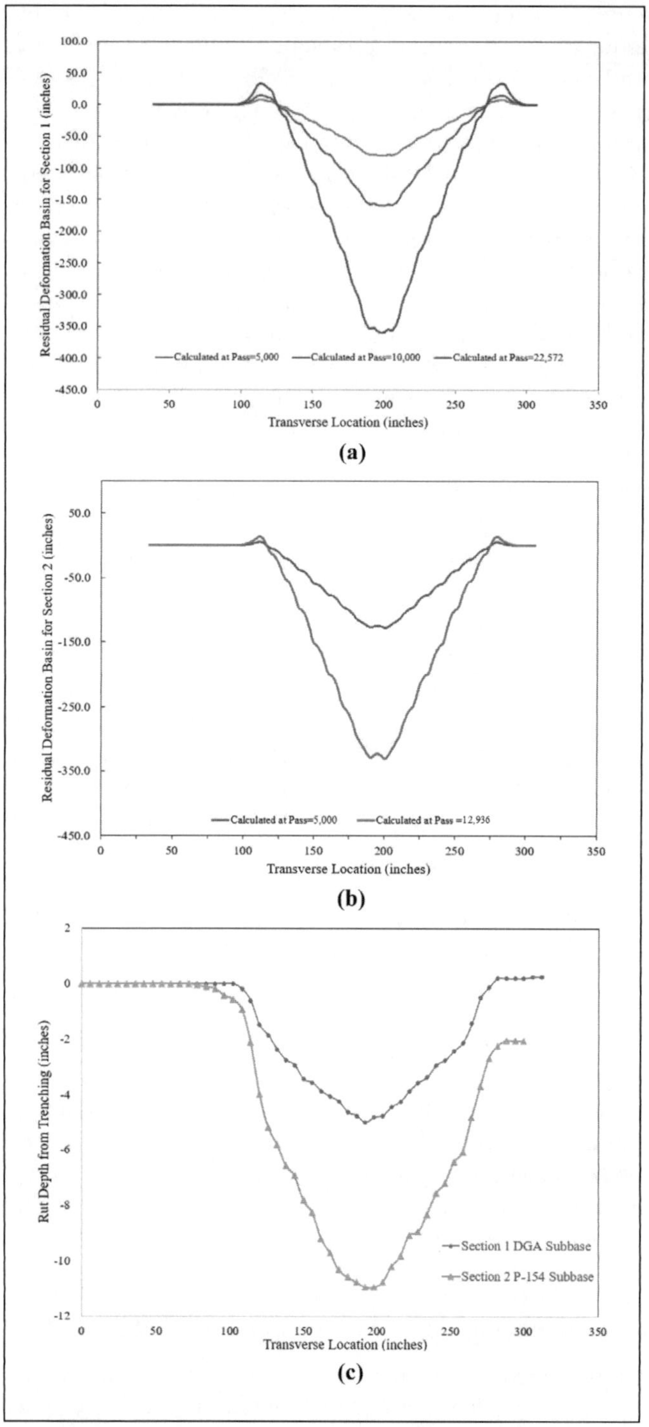

Figure 9. (*a*) Residual deformation basin for Section 1; (*b*) residual deformation basin for Section 2; (c) rut depths calculated from trenching (1 inch=25.4 mm).

aircraft movement and the measured transverse profile after a complete wander pattern (*14*). The CC1 and CC3 study found that unique relationship exists between the maximum downward residual deformation and the critical transverse points. It was also reported that a closer deformation profile could be developed if the calculation of the transverse profile was corrected by considering the residual deformation of the previous pass. The stress-history-based approach considers the fact that if there was rut due to previous pass and there should be rut in same location due to the current pass as well, the additional rut would be less. Similarly, if there was heave in the previous pass and there should be heave in the next pass but with less additional heave.

Stress-history correction was applied to the initially computed surface profile (as previously shown in Figure 9*a* and *b*) by assuming that if the previous pass caused the same type of residual deformation as the current pass, then the residual deformation by the current pass was reduced by multiplying the current pass residual deformation value by a constant reduction factor. During this study, the constant reduction factors were always the residual values as observed at the critical locations for the return pass i.e., pass that took place at east to west direction (wander position 4 and −3).

Figure 10*a* and *b* show the calculated surface transverse profile for Section 1 and 2, respectively, with correction factors applied for the previous pass. As shown in these figures, by considering the stress-history effects, the magnitude of the calculated residual deformation reduced significantly and were much closer to the measured values. The jagged nature of the graphs was the result of the rough estimation of the effects of the previous pass. Though the magnitude of the calculated rut profile reduced significantly after the constant reduction factor was applied, the presented graphs indicate improvements to this method are still needed. One of the limitations of the model was that the surface profile was extrapolated linearly from the 66-pass residual basin. The predicted profile would have matched the nature of the field rut crater more closely if the final profile was predicted using graduated influence factors calculated based on the transverse surface profiles measured in the field. However, since field transverse profiles were not available, it was near impossible to calculate the graduated influence factor based on the magnitude of the difference between the previous and current pass residual deformation values. And in the absence of that information, the current model could not be refined further. Regardless, using the influence of the previous pass on the current one is promising. The advantage to using the stress-history correction procedure is that any combination of wander positions can be simulated to predict the future deformation basin.

Summary and Conclusion

An attempt was made to predict the rut depths in airport pavements due to realistic air traffic using the MDD data

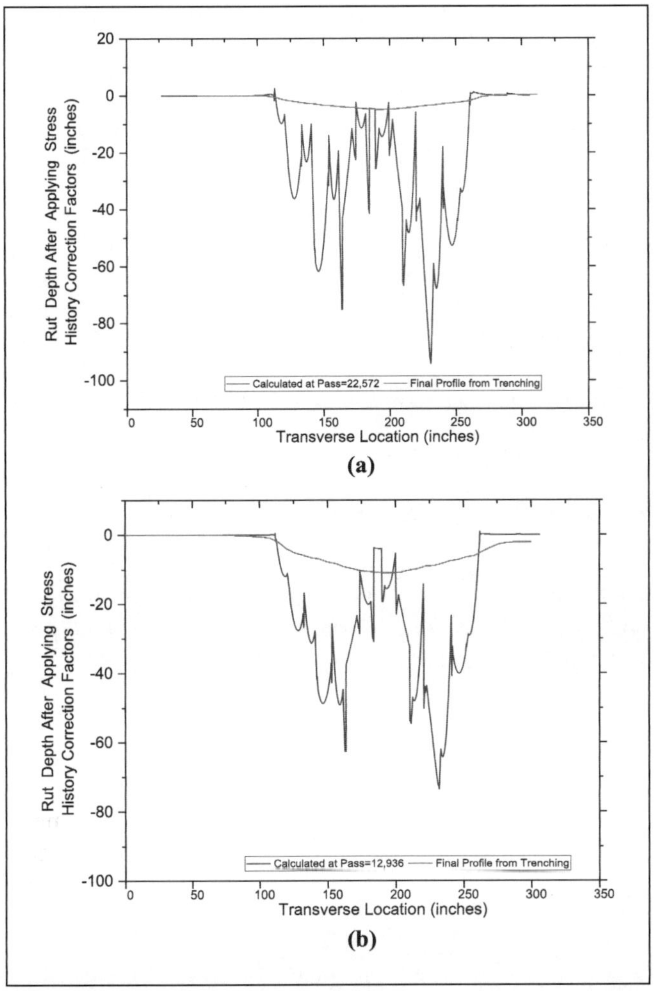

Figure 10. Calculated rut depths corrected for stress history for (*a*) Section 1 and (*b*) Section 2 (1 inch=25.4 mm).

from the CC5 test sections built at the FAA NAPTF. A method of using only the critical locations was proposed to develop a transverse profile for each pass and calculate the transverse profile created by multiple passes. This method was based on the relationship between the maximum residual MDD readings due to various wander positions. Using the measured transverse profile from a single 66-pass wander pattern, it was possible to create a theoretical individual pass transverse profile that when combined over the 66-pass wander pattern produced a residual deformation basin similar to the contour of the field measured profile.

However, the initial attempt at using this method did not consider previous load applications when determining the rut or heave of the current pass and thus the rut depths were larger than the measured values. By considering the stress-history effects from the previous pass, it was possible to produce a more accurate residual deformation profile with reduced rut depths.

Further refinement of the proposed methodology using field measured transverse surface profiles has the potential to estimate realistic permanent deformations of future pavement test section by simply collecting residual deformation data from a few hundred initial trafficking load cycles. And the ability to predict the future rut profiles will ensure an accurate estimation of the remaining life of pavements, helping the designers plan an effective maintenance and rehabilitation strategy, subsequently leading to an economic and sustainable airport pavement management system.

Acknowledgments

This study was sponsored by the Airport Cooperative Research Program through a Graduate Research Award. The authors acknowledge Lawrence Goldstein and Dr. Dominque Pittenger for their guidance and investment in this research. Authors are especially grateful to Dr. Navneet Garg of Federal Aviation Administration (FAA) for his valuable feedback and insightful direction to the individual research tasks.

Author Contributions

The authors confirm contribution to the paper as follows: study conception and design: PS and ET; data collection, analysis, and interpretation of results: PS; draft manuscript preparation: PS and ET. All authors reviewed the results and approved the final version of the manuscript.

References

1. Ho Sang, V. A. *Field Survey and Analysis of Aircraft Distribution on Airport Pavements.* Report No. FAA-RD-74-36. U.S. Federal Aviation Administration, Washington, D.C., 1975.
2. Kim, I. T., and E. Tutumluer. *Permanent Deformation Behavior of Airport Flexible Pavement Base and Subbase Courses.* FAA Center of Excellence for Airport Technology (CEAT) Report No. 28. Department of Civil and Environmental Engineering, University of Illinois at Urbana-Champaign, Champaign, IL, 2005.
3. Donovan, P. R., and E. Tutumluer. Analysis of NAPTF Trafficking Response Data for Pavement Foundation Deformation Behavior. *Proc., 2007 Worldwide Airport Technology Transfer Conference*, Atlantic City, NJ, Federal Aviation Administration, Washington, D.C., 2007.
4. Donovan, P. R., and E. Tutumluer. The Anti-Shakedown Effect. *Proc., 1st International Conference on Transportation Geotechnics*, Nottingham, UK, 2008.
5. Donovan, P. R., and E. Tutumluer. Effect of Aircraft Load Wander on Unbound Aggregate Pavement Layer Stiffness and Deformation Behavior. *Proc., Transportation and Development Institute of the American Society of Civil Engineers Airfield and Highway Pavements Specialty Conference*, Bellevue, WA, 2008.
6. Donovan, P. R. *Analysis of Unbound Aggregate Layer Deformation Behavior from Full Scale Aircraft Gear Loading with Wander.* PhD dissertation. University of Illinois at Urbana-Champaign, Champaign, IL, 2009.

7. Federal Aviation Administration. *Airport Pavement Design and Evaluation*. AC 150/5320-6E. FAA, U.S. Department of Transportation, Washington, D.C., 2009.
8. Hayhoe, G. F., and N. Garg. Posttraffic Testing on Medium-Strength Subgrade Flexible Pavements at the National Airport Pavement Test Facility. *Proc., ASCE Airfield Pavement Specialty Conference*, Las Vegas, NV, American Society of Civil Engineers, Reston, VA, 2003, pp. 136–146.
9. Garg, N., and G. F. Hayhoe. Permanent Deformation Behavior of the Granular Layers Tested at the FAA's National Airport Pavement Test Facility. *Proc., APT'08 3rd International Conference*, Madrid, Spain, 2008.
10. Garg, N., F. Pecht, and Q. Jia. Subgrade Stress Measurements under Heavy Aircraft Gear Loading at FAA National Airport Pavement Test Facility. *Proc., Geo-Shanghai 2010*, ASCE Special Publication No. 203, China, 2010, American Society of Civil Engineers, Reston, VA, pp. 484–491.
11. Hayhoe, G. F. Traffic Testing Results from the FAA's National Airport Pavement Test Facility. *Proc., 2nd International Conference on Accelerated Pavement Testing*, University of Minneapolis, MN, 2004.
12. Hayhoe, G. F., N. Garg, and M. Dong. Permanent Deformations during Traffic Tests on Flexible Pavements at the National Airport Pavement Test Facility. *Proc., 2003 Airfield Pavement Specialty Conference (APSC)*, Las Vegas, NV, 2003, Airfield Pavements: Challenges and New Technologies, 2004, pp. 147–169.
13. Werkmeister, S., A. Dawson, and F. Wellner. Permanent Deformation Behavior of Granular Materials and the Shakedown Concept. *Transportation Research Record: Journal of the Transportation Research Board*, 2001. 1757: 75–81.
14. Donovan, P., P. Sarker, and E. Tutumluer. Rutting Prediction in Airport Pavement Granular Base/Subbase: A Stress History Based Approach. *Transportation Geotechnics*, Vol. 9, 2016, pp. 139–160.

The ACRP Selection Panel for the Graduate Research Award Program on Public-Sector Aviation Issues peer-reviewed this paper (18-03086).

The contents of this paper reflect the views of the authors who are responsible for the facts and the accuracy of the data presented. The contents do not necessarily reflect the official views or policies of the Federal Aviation Administration or the Illinois Center for Transportation. This paper does not constitute a standard, specification, or regulation.

Article

TRR
JOURNAL OF THE TRANSPORTATION RESEARCH BOARD

Transportation Research Record
2018, Vol. 2672(29) 93–102

DOI: 10.1177/0361198118799709
journals.sagepub.com/home/trr
SAGE

Germs on a Plane: The Transmission and Risks of Airplane-Borne Diseases

Nereyda L. Sevilla[1]

Abstract

This research explored the role of air travel in the spread of infectious diseases, specifically severe acute respiratory syndrome (SARS), H1N1, Ebola, and pneumonic plague. Air travel provides the means for such diseases to spread internationally at extraordinary rates because infected passengers jump from coast to coast and continent to continent within hours. Outbreaks of diseases that spread from person to person test the effectiveness of current public health responses. This research used a mixed methods approach, including use of the Spatiotemporal Epidemiological Modeler, to model the spread of diseases, evaluate the impact of air travel on disease spread, and analyze the effectiveness of different public health strategies and travel policies. Modeling showed that the spread of Ebola and pneumonic plague is minimal and should not be a major air travel concern if an individual becomes infected. H1N1 and SARS have higher infection rates and air travel will facilitate the spread of disease nationally and internationally. To contain the spread of infectious diseases, aviation and public health authorities should establish tailored preventive measures at airports, capture contact information for ticketed passengers, expand the definition of "close contact," and conduct widespread educational programs. The measures will put in place a foundation for containing the spread of infectious diseases via air travel and minimize the panic and economic consequences that may occur during an outbreak.

The risk of international spread of diseases has increased because of the advances in technology that have made global air travel a daily occurrence. More than 4.5 billion individual air journeys are made every year, increasing the risk of disease spread beyond the local area as passengers travel from coast to coast and continent to continent (Figure 1) (*1, 2*).

In the twenty-first century, air travel provides new means for diseases to spread internationally at extraordinary rates because people have the ability to jump from country to country in hours. This was evident in the 2003 severe acute respiratory syndrome (SARS) pandemic that killed over 800 people across 37 countries and the 2009 influenza H1N1 epidemic that affected over 200 million individuals. By the time travel alerts and health recommendations were announced, the diseases had already reached rampant proportions, costing thousands of lives and billions of dollars. It is incumbent on scientists to develop tools that will track and predict disease spread and identify interventions that can be executed in a timely and effective manner. Plans must be based on technical and scientific knowledge of the vectors involved and the characteristics of the emerging infectious diseases.

Methodology

This research used a mixed methods approach to evaluate the impact of air travel on the spread of infectious diseases and the effectiveness of public health strategies to mitigate disease outbreaks. A tool known as the Spatiotemporal Epidemiological Modeler (STEM), developed by IBM as an open-source program, was used to simulate four disease outbreaks in the United States, specifically SARS, H1N1, Ebola, and pneumonic plague (Table 1). Table 1 depicts the biological characteristics of the four diseases. It is important to note that SARS, H1N1, and pneumonic plague can all spread from person to person by air from coughing. Ebola is spread by the transfer of bodily fluids from person to person (e.g., sex). Another point of significance is that although pneumonic plague is extremely deadly—it kills a person in one to three days—opportunities for contagion are, therefore, limited. SARS and H1N1 have lower mortality

[1]George Mason University, Fairfax, VA

Corresponding Author:
Address correspondence to Nereyda L. Sevilla: nsevilla@gmu.edu; nereydasevilla@msn.com

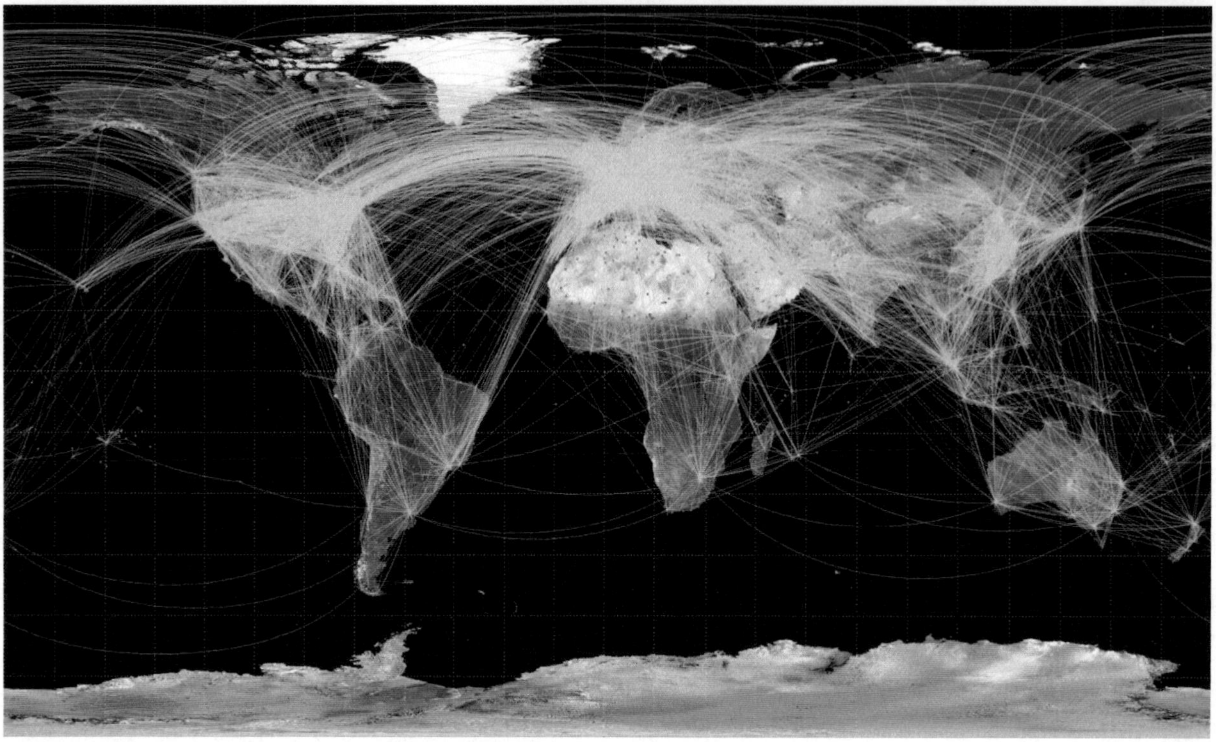

Figure 1. Global air travel routes.

rates, but a contagious person can spread either disease for several weeks.

STEM is a multidisciplinary, collaborative modeling platform that uses compartment theory to simulate the spread of disease (*3*). The open-source characteristics of the system allow researchers to compare, refine, and validate different scenarios as well as add data such as disease characteristics. STEM provides the built-in statistics such as county and country boundaries, transportation networks, air travel information, and environmental conditions. The equations used within the compartment model provided the foundation for the model that would be used to compare the threat of disease spread while holding many characteristics constant. In compartment models, each person in a population is accounted for in a compartment, and no one person may be in more than one compartment at any given time. All the diseases of interest for this research used an "S"– "E"– "I"– "R" (SEIR) model, in which each person in a population is in one of four states: susceptible (S), exposed (E), infectious (I), or recovered (R) (Figure 2). The disease models can be configured with known disease characteristics, including infection rates, incubation period, and mortality rates. These variables can also be changed during the course of the scenario to depict possible mutations. Transitions between compartments are driven by differential equations (Figure 2). The vectors of disease that affect the transmission of an infectious disease, such as motion of individuals, ground travel, air travel, and social gatherings, can also be modeled.

The simulations in this research forecast the natural flow of disease with air travel starting with one or 10 infected individuals traveling from New York City. By applying the lessons of real-world outbreaks, public health officials can make informed decisions as to the best interventions and communication methods should a disease outbreak occur.

Results

The results illustrate the impacts of hypothetical outbreaks in the United States, starting in New York, beginning in September and lasting for 6 months. These simulated cases of SARS, H1N1, Ebola, and pneumonic plague are all based on known infection parameters, such as basic reproductive ratio (R_o), transmission rate (β), incubation rate (ε), recovery rate (γ), and infectious mortality rate (δ), and are based, in turn, on data from previous outbreaks, historical epidemiological numbers, scientific evidence, and biological characteristics as outlined in Table 1.

The models of SARS, H1N1, Ebola, and pneumonic plague mirror the actual historical outbreaks in the

Table 1. Biological Characteristics of Disease of Interest

	SARS	H1N1	Ebola	Pneumonic plague
Family	Coronavirus (virus)	Orthomyxoviridae (virus)	Filoviridae (virus)	Enterobacteriaceae (bacteria)
Year of outbreak	2003	2009	2014	Hypothetical
Method of transmission	Person to person by air (e.g., coughing) and fecal–oral transmission (e.g., putting dirty hands near mouth)	Person to person by air (e.g., coughing)	Direct person-to-person contact through bodily fluids (e.g., sex)	Person to person by air (e.g., coughing)
Asymptomatic transmission possible?	Low possibility	Yes	No	No
Reproductive ratio (average # of new cases generated by each case) – R_o	2–3	1.4–3.5	1.5–2.0	1.3
Case fatality rate	13% $<$60 yrs 43% $>$60 yrs	0.01%–0.3%	50%–90%	50% (if untreated, 99%)
Symptoms	Fever, general flu-like symptoms, muscle pain	Sudden fever, body aches (joints and throat), coughing, sneezing, extreme chills, fatigue, nasal congestion	Severe frontal and temporal headache, aches and pains, fever progressing to watery diarrhea, abdominal pain, nausea, vomiting	Sudden headaches, chills, malaise, and increased respiratory and heart rates progressing to cough and fever
Treatment or vaccine	Oseltamivir, supportive care, experimental vaccine	Oseltamivir, flu shot	Supportive care, experimental vaccines	Antibiotics effective if within 24 hours of symptoms
Incubation period	1–14 days	2–6 days	2–21 days	2–4 days
Duration of illness	2–4 weeks	1–2 weeks	10–20 days	Death if untreated in 1–3 days

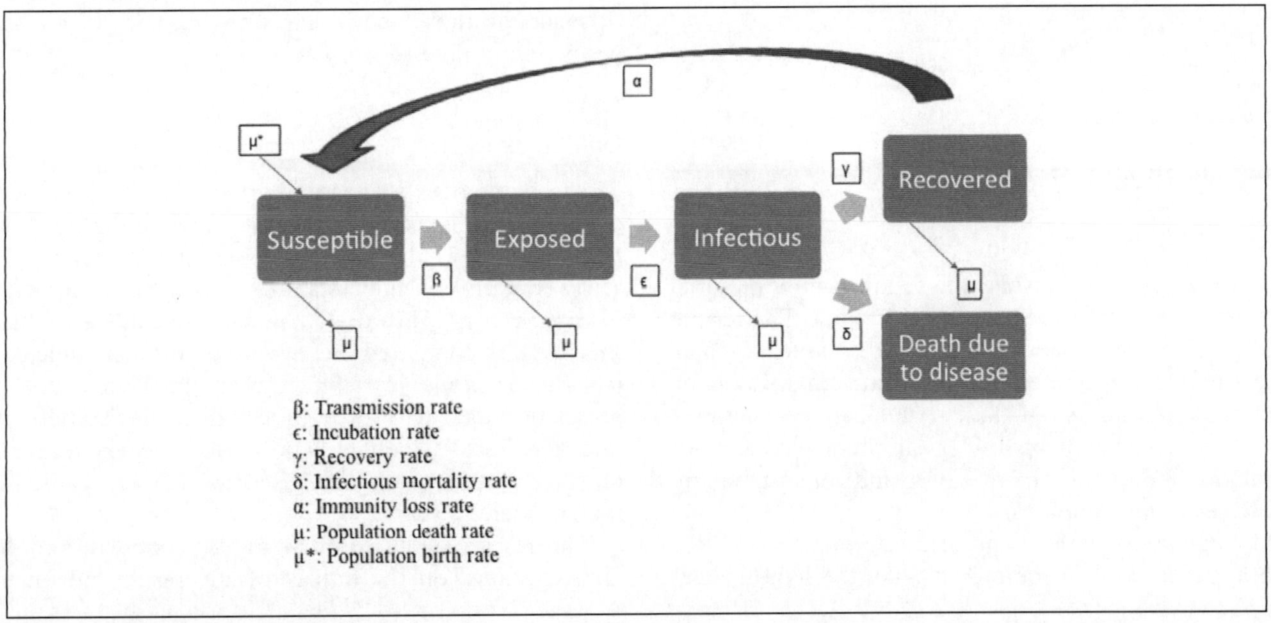

Figure 2. SEIR compartment model.

Table 2. Simulated and Actual Disease Cases

Simulated and actual cases in the United States		
Simulated diseases (1 or 10 initial cases)	Total infected (% of the population)	Total deaths
SARS – 1	4,042 0.0014%	572 0.0002%
SARS – 10	7,279 0.0026%	1,000 0.0004%
H1N1 – 1	5,617,702 1.9962%	5,374 0.0019%
H1N1 – 10	18,363,550 6.5253%	18,856 0.0067%
Ebola – 1	5 <0.0001%	1 <0.0001%
Ebola – 10	43 <0.0001%	19 <0.0001%
pneumonic plague – 1	1 <0.0001%	1 <0.0001%
pneumonic plague – 10	25 <0.0001%	13 <0.0001%
Historical diseases	Total infected	Total deaths
2003 SARS United States	27 <0.0001%	0 0%
2003 SARS Canada	438 0.0015%	44 0.0001%
2003 SARS China	5,327 0.0004%	349 <0.0001%
2009 H1N1 United States (Apr–Oct 2009)	~22 million 7.8174%	~3,900 0.0014%
2014 Ebola United States	4 <0.0001%	1 <0.0001%
2015 pneumonic plague United States	11 <0.0001%	3 <0.0001%

United States and are assumed to follow similar patterns if another outbreak were to occur (Table 2). The results indicate that SARS and H1N1 have a much greater impact in relation to infections and deaths than Ebola or pneumonic plague. All of these results assume a natural flow of disease based on biological characteristics and population movement and do not include or assume any public health interventions.

In comparison with the historical accounts of SARS, H1N1, Ebola, and pneumonic plague, the hypothetical scenarios of each of the disease spreads follow similar patterns and results. The model shows that an outbreak of SARS in the United States would cause approximately four to seven thousand cases and 500 to 1,000 deaths. Although the United States experienced a small number of SARS cases but no SARS deaths, its neighboring country of Canada did confront 438 cases and 44 deaths (*4*). Of the four diseases, H1N1 has the most impact on the population according to the model, with 5 to 18 million people affected and five to 18 thousand deaths. During the 2009 H1N1 outbreak, the United States experienced approximately 22 million disease cases and approximately 4,000 deaths during a 6-month period (*5*). During the Ebola crisis in 2014, the United States had two imported cases of Ebola, and two locally acquired cases by medical professionals. Specifically two nurses that directly treated the first case of imported Ebola were the only locally transmitted cases of Ebola in the US. No cases of Ebola were acquired through air travel. Finally, the United States experiences approximately 10 to 15 cases of pneumonic plague per year, but has not had a person-to-person spread of pneumonic plague since 1925 (*6*). There has never been a case of pneumonic plague associated with air travel. The model shows very small numbers of individuals infected by and dying because of pneumonic plague and Ebola as compared to SARS or H1N1 should outbreaks occur.

The comparative results of each of the four modeled diseases along with the historical accounts show the importance of the disease characteristics and the impact of the infection rate. A disease that has a long period of illness, such as SARS or H1N1, is expected to cause a higher natural spread than diseases in which the period of illness is brief. In cases in which the period of illness is short, such as pneumonic plague, those individuals affected do not have the same opportunities to infect others as do those people harboring diseases with a longer period of illness.

Limitations

The technology exists to rapidly build many new models of infectious diseases. A "good approximation" requires not only knowledge of the disease parameters themselves, but also an understanding of the most important disease vectors. The SEIR compartment model used in this research compared one disease to another while controlling for similar variables such as environmental and population data. However, in actuality, the reproductive rate of a disease, the incubation rate, recovery rate, and mortality rate can all vary based on socioeconomic factors, gender, and age.

This research focused on an average population in the United States, on the impact of air travel, and on the policy for controlling the spread of disease, based on typical disease characteristics. For air travelers, exposure is often related to their purpose for travel, that is, visiting friends and interacting with the population, or pure tourist or business travel, in which exposure may be less. Infectious individuals in the model travel at the same rate as non-infectious individuals, which may not be true under real travel conditions (*7*). Sick individuals may elect to stay home and this would reduce the rate of infection and help minimize the disease outbreak. However, the model does normalize air travel for individual diseases and allows for the long-distance jump from one coast to another (*8*). These limitations in the overall computer modeling are minor because the model provided for control of travel among the different diseases depicted. The outcomes are seen in the trends as one disease is compared to another, but the numbers of infectious or deceased individuals should not be used as absolutes.

Furthermore, the models assume the pathogens of interest mirror previous outbreaks with known disease features. If the genetics of any of the diseases are changed by natural mutation or by intentional genetic manipulation, the outcomes of disease spread may change dramatically. The manipulation of the disease characteristics may change the recommendations if the disease spread or mortality rate is greater than previously observed. This research assumes no manipulation or genetic mutation has occurred.

Recommendations

In the event of an infectious disease epidemic, the world must act swiftly and decisively, and unite to form a cohesive plan based on science to limit the spread of disease, thus reducing economic costs and psychological impacts. Officials must implement measures to track the pattern of disease spread as soon as the threat is identified, especially if the disease of concern, for example SARS or H1N1, has a high transmission rate. Therefore, when an infectious disease outbreak occurs, it is imperative to have appropriate travel and communication policies in place. In cases involving Ebola and pneumonic plague, efficient and cost-effective measures will limit apprehension and minimize the economic consequences.

This research outlined physical and policy recommendations for implementing effective measures in response to a disease outbreak; all are rooted in the education of the public. Mitigation measures are also outlined in the *Airport Cooperative Research Program Report 91: Infectious Disease Mitigation in Airports and on Aircraft*, which includes 24 recommended actions for buildings, airplanes, and people (*9*). Many of the recommendations in this report are reiterated in this research; six specific recommendations are expanded upon below. Research showed that simple measures may improve global travel health. Their implementation has the potential to limit death, minimize infections, decrease economic impacts, and curtail fear.

Recommendation #1: Expand the Definition of "Close Contact"

The World Health Organization (WHO) defines close contact as the same row plus two rows ahead and two rows behind the identified ill individual (*10*). Previous research indicated that the risk of transmission while traveling in an airplane is very low. (*10–13*) An airplane has a protective mechanism during flight that includes air being constantly circulated through high-efficiency particulate air filters and mixed with outside air. However, close contact with ill individuals in airports may prove to be a more likely route of transmission than contact on the airplane. Airport scenarios include and are not limited to: extended close contact while seated prior to boarding; close contact with individuals at neighboring gates traveling to other destinations; delays that necessitate being near to other travelers in lines at check-in, security screening, restrooms, and concessions; and close contact in a confined jetway space while waiting to board. Because individuals are encouraged to arrive at the airport hours before a flight, they may spend more time at the airport than on the airplane itself. During this time, an individual may be in close contact with many more individuals outside of an airplane than while flying. In such cases, non-infected individuals may be exposed to infectious individuals. Policies must take into account all the individuals on an airplane and within the timeframe of the exposure window at the airport. It will take the cooperation of public health officials and government agencies to recognize the science of disease spread and respect the privacy of individuals and, therefore, communicate to the traveling population in the most appropriate way the possible risk of disease infection, thus minimizing its spread.

Recommendation #2: Health Contact Information Requirement on all Air Travel Ticket Purchases

The use of a passenger manifest to track down and trace passengers who may have been affected by a diseased individual is not only very resource intensive but it may also miss a large population of other close contacts outside of the airplane. To improve contact information across airport populations, additional contact requests should be required on all tickets (Table 3).

Efficient capture of passenger contact information is a necessary response to a disease outbreak. If the ticketed passengers at the airport have already given their contact information, a widespread communication effort can be

Table 3. Example of Disease Notice on Air Travel Ticket Purchases

Notification of disease in travel areas, airports, or on airplanes
Do you wish to be notified of an outbreak of any infectious disease of concern* reported in the travel area, airport, or on board the airplane? ■ Yes ■ No
If yes, contact information must be provided: ■ Text: ■ E-mail: ■ Voicemail: ■ Other, please provide complete contact information:
All current travel alerts may be found at: http://wwwnc.cdc.gov/travel/notices/.
*Infectious diseases of concern include: (1) cholera, (2) diphtheria, (3) infectious tuberculosis, (4) plague, (5) smallpox, (6) yellow fever, (7), viral hemorrhagic fevers, (8) SARS, and (9) flu that can cause a pandemic.
Please consult the CDC website to determine if an outbreak is relevant to your travels.

Table 4. Example of a Health Notice Given during Air Travel Ticket Purchase

WARNING: TRAVEL AREA HAS HEALTH ALERT FOR PNEUMONIC PLAGUE!
All current disease travel alerts and advisory notices may be found at: http://wwwnc.cdc.gov/travel/notices/.
Do not travel if suspected of carrying a disease of concern.* *Infectious diseases of concern include: (1) cholera, (2) diphtheria, (3) infectious tuberculosis, (4) plague, (5) smallpox, 6) yellow fever, (7), viral hemorrhagic fevers, (8) SARS, and 9) flu that can cause a pandemic.
Please consult the CDC website to determine if an outbreak is relevant to your travels.

made to reach all those with tickets at the airport during the time of the infectious exposure.

Recommendation #3: Expand Passenger Airport and Pre-Boarding Self-Sanitizing Measures

Filters on airplanes serve as a baseline preventive measure against in-flight disease transmission. However, their effectiveness is greatly increased when passengers themselves practice infection control techniques. Airport announcements and visual aids about disease preventive measures displayed around the terminal provide instant reminders and education. This just-in-time education would be fresh in passengers' minds and emphasize to them that they will be at close quarters with other travelers at the gate, on the jetway, during the boarding process, and during flight. Airlines and airports should make hand sanitizers readily available, especially at restroom entrances, at the gate area before boarding an airplane, at all food court locations, at the entrances of all the airport stores, in all the lounge areas, and randomly around all the corridors. During embarkation, the gate officials should be encouraging everybody to use hand sanitizers and expand personal space on the jetway. These immediate instructions would increase compliance. At the same time, airport announcements should remind passengers to cover their coughs with hands or elbows. Additionally, airlines could provide personal handwipes, allowing passengers to clean their hands and other areas that harbor germs, such as fold-down trays. Everybody should carry out the preventive measures. This continuous emphasis on good habits may start a culture trend and natural habit formation, not just during an outbreak, but at all times, and such low-cost preventive measures may minimize pathogen transmission.

Recommendation #4: Enhanced Travel Alerts and Advisory Notices during Ticket Sales

The use of health statements on air travel ticket purchases is also a way of encouraging every passenger to curtail personal travel to world areas of concern as well as limit personal travel during an illness. Travel restrictions have been found to be minimally effective and must be strictly enforced to be successful. Historical data have indicated that travel restrictions may only cause a slight delay in infectious disease introduction to the United

Table 5. Example of a Thermal Screening Alert on an Air Travel Ticket

ALERT **SARS Outbreak Concern in New York City** **Random Thermal Screening Instituted**
Your travel plans involve an airplane flying from New York City.
You may be subjected to random thermal screening due to a suspected outbreak of SARS. If you suspect you are ill, do not travel and delay your travel until fever symptoms have subsided. Please consult a doctor if you have concerns about disease exposure. Do you understand that you may undergo thermal testing during your travels? ■ Yes ■ No
Please consult the CDC website for more travel information: http://wwwnc.cdc.gov/travel/notices/.

States (*10*, *14*, *15*). It is more cost effective to inform the public about the various appropriate resources available in order for them to make a more informed choice. Notices of health concerns should be displayed during online ticket purchase or at a ticket counter and relayed over the phone (Table 4).

Travel alerts and advisory notices in combination with educating travelers about proper preventive measures would reduce the risk of infection. The advisory notices should come from major health agencies such as the WHO and the Center for Disease Control and Prevention (CDC). A strong message may encourage the public to further find information about a disease and their risk of infection. In the event of an infectious disease outbreak, clear and concise travel alerts will help in reducing the spread of infection through air travel. Each of the agencies should declare a notice when scientifically applicable and disseminate that information through various social media outlets, thereby enhancing those messages given at the airport and accompanying each ticket sale.

Recommendation #5: Limited, Announced, Random Temperature Checks

Targeted entry and exit procedures have been shown to have a limited effect on the spread of disease (*16*, *17*). However, in a dire situation, these very costly procedures may be a measure of last resort to prevent a deadly disease from spreading. It must be noted, however, that the deadliness of the disease must outweigh the economic and political impact of such a decision. It would also only be effective with the complete cooperation of all international airlines and country public health officials.

Limited and announced, yet random temperature checks during an outbreak may deter ill individuals from traveling. Thermal screenings themselves may not be cost effective in halting the spread of disease. However, the threat of entry and exit procedures may be sufficient to deter as many individuals from traveling as would a large-scale thermal screening effort. If an infectious disease such as SARS is detected in New York, then any airplane from New York airports may be subjected to random screening tests. If, during the ticketing process, a passenger is alerted to the random check, that person may decide not to purchase the ticket or change their travel plans (Table 5).

Similarly, during check-in, if an outbreak is confirmed in a travel area, the government and airlines should work together to ensure passengers are incentivized to rethink their travel plans. These measures include a full refund or changes in dates without penalty. The same should be true if passengers find themselves ill. To prevent the spread of disease, passengers should be allowed to change their plans without penalty after consultation with and providing documentation from authorized physicians. Working with public health officials, members of the public may take personal responsibility and postpone travel until the threat of contagion has passed.

Recommendation #6: Specific Crisis Communication

The basis for early containment procedures is constant and relevant communication. Early messages during a disease crisis must educate the public on the medical threats during travel. Education of the public is the crucial foundation for and probably the most cost-efficient and effective way of preventing the spread of disease in the first place and slowing the spread if it should start.

In the event of an outbreak, communication barriers arise when individuals take advice from a source of their choosing, especially if that source is not an expert. Emergency response officials must provide consistent, accurate, and simple information that can help the public prevent transmission of infectious disease, detect symptoms, and seek treatment so that uncertain individuals do not look elsewhere for information. Messages should provide pertinent and applicable information such as general prevention methods, the differences in the presentations of various illness, and treatment methods.

Table 6. Example of a Pneumonic Plague Message Map

Stakeholder: Public
Question: Will I get pneumonic plague on an airplane when flying to or from an outbreak area?

Key message 1	Key message 2	Key message 3
Pneumonic plague is extremely difficult to contract person to person	Infection rate is extremely low	It is very difficult to spread pneumonic plague coast to coast
Supporting information 1–1	Supporting information 2–1	Supporting information 3–1
85% of pneumonic plague (also known as bubonic plague) cases are caused by transmission of the disease by rodents, and the disease does not spread easily from human to human	A person is only infectious for 1–3 days	The United States reports approximately 25 cases of pneumonic plague a year. There has not been a sustained outbreak for nearly a century.
Supporting information 1–2	Supporting information 2–2	Supporting information 3–2
Pneumonic plague spread from person to person usually occurs after long periods of close contact, for example, in relation to caregivers or medical professionals	A person becomes ill very rapidly and it would be difficult to travel while experiencing symptoms	A disease with such a low infection rate cannot sustain a US-wide outbreak
Supporting information 1–3	Supporting information 2–3	Supporting information 3–3
A person is not infectious without symptoms	A person suffering from pneumonic plague would be easier to spot than someone suffering from another disease due to the plague's rapid progression and severity	Previous large numbers of pneumonic plague deaths were due to unsanitary conditions, lack of modern medicine and antibiotics, and large numbers of rodents

To guide public health officials, the WHO produced a handbook entitled *Effective Media Communications during Public Health Emergencies* (*18*). This has been the guiding foundation of many instructions issued by countries and local organizations in the midst of a crisis. Critical tips for effective communication include:

Provide Early and Constant Communication and Avoid Rumors from Non-Credible Sources. Individuals will start doing their own research when the rumor of a disease outbreak begins to circulate. Efforts should be made to ensure credible websites and information sources are prominently advertised. Individuals should be steered toward websites such as CDC Travelers' Health (https://wwwnc.cdc.gov/travel), U.S. Department of State (sections Passports, and International Travel) (https://travel.state.gov/content/travel/en.html), and WHO International Travel and Health (http://www.who.int/ith/diseases/en/). Potential passengers may also start looking at travel websites, airline home pages, or other sources of information, official or unofficial. The timeliness with regard to how and where messages are conveyed may help lessen panic. However, studies have shown that under stress, individuals usually compare current messages to the first pieces of information processed, even if later messages are more accurate (*19*). It is imperative that scientifically validated messages are posted in multiple locations, especially on common travel websites and in news outlets.

Simple, Honest Communication Statements Created from Message Maps. The CDC recommends constructing message maps to convey standardized messages that target specific audiences. A message map packages important facts about the disease and the risks of disease spread in simple sentences. Public relations offices could use message maps to guide their announcements so that honest, trustworthy, and relevant information can be conveyed to potential travelers. As an example, because pneumonic plague already has a fearsome reputation as the "Black Death," it is essential to provide accurate and easily accessible information in a message map (Table 6).

In the event of a disease outbreak, it is as important to state what is unknown as it is to state what is actually known. Public health and government officials maintain credibility when they explain what factors are still unknown instead of stating what the public may want to hear. More panic may be created if a wrong message is conveyed. Even if officials present new and accurate information later, their credibility could have been destroyed by not being honest earlier about the unknowns. Therefore, it is best that public officials state "We don't know," because an educated guess may backfire in the long run. An untrue statement will become a roadblock to any future truthful or useful statements. All statements need to be verified by scientific evidence from validated sources.

Furthermore, it is also important that officials advise the public when further information will be available, how those messages will be presented, and when official social media and websites will be updated. The public will look for new information if told when and where to find it in its most up-to-date form. However, it is key not to give false hope or guarantee victory unless a victory is 100% guaranteed. If the source of the disease has been contained and secondary infections isolated, then the public can be reassured.

Inaccurate information, and promises broken early on may worsen a crisis. The public will lose all trust if early in a crisis the officials prove to be wrong and may turn to anyone who appears to be right even though those individuals are unqualified. Public officials need to establish their expertise and credibility as soon as possible.

Empower the Public. The public must be shown how their actions will help prevent the spread of disease. Educational programs should include: the necessity of coughing into a hand/elbow during a flight and the use of hand sanitizer. If officials lead by example, it may create a natural habit pattern. When actions of trusted individuals mirror the health guidelines, the public will be more accepting of the underlining message. Studies have shown that if the public can take actions during a crisis it can help restore a sense of control and overcome feelings of hopelessness and helplessness (*19*).

Different media outlets should disseminate the same messages to accommodate the various generational, political, and social differences and, particularly, these should be tailored to their local population. Despite the differences in media markets, all should broadcast the same consistent message. National and local leaders, including educators and religious officials, can also disseminate simple preventive and basic infection control measures. For example, teachers can remind students to wash their hands before lunch periods. These simple messages will have a lasting impact, reduce stress and, ultimately, lessen the spread of disease.

Discussion of Results

The spread of disease during travel is a concern; however, diseases such as SARS and H1N1, which have a high transmission rate, will have a greater impact on the population more in relation to infections and deaths than Ebola or pneumonic plague, even though the latter two may present more of a psychological panic. Efforts should be made to contain the disease and communicate the threat based on scientific evidence. The airport itself may be more of a location for transmission of disease because individuals wait in the same ticketing lines, gates, and jetways as infected passengers. The recommendations presented are relevant to passengers who are in the vicinity of infected individuals at the airport and not just traveling on an airplane.

Historical examples and models show that the spread of SARS and H1N1 is significantly affected by air travel. Because of the advancements in travel technology and engineering, humans can reach any part of the globe within the incubation time of most diseases, allowing travelers to unintentionally carry a disease. Ebola and pneumonic plague have short illness durations with a high probability of death, so the risk of transmission is low and the outbreak is not sustainable without further incubators. However, all public health professionals should be prepared to deal with a disease that may not be endemic to an area. Because of air travel, all areas of the Earth are susceptible to all types of diseases.

In each of the recommendations, the component of public education is of the utmost importance. The scientific and medical advancements in disease prevention and treatments mean that many infectious diseases are no longer seen as serious biological threats. However, many diseases, such as pneumonic plague, remain as serious psychological threats. Public health officials need dedicated resources with which to create communication strategies to reduce the psychological fallout from a disease outbreak. This, in turn, minimizes the economic consequences from such an outbreak.

To contain the outbreak of diseases, aviation and public health authorities should establish preventive infectious disease measures at airports, capture contact information for ticketed passengers, expand the definition of "close contact," and conduct widespread educational programs. These measures will put in place a foundation for containing the spread of infectious diseases, minimize the panic, and reduce costs that may occur during an outbreak.

Acknowledgments

STEM is an Eclipse project originally created by IBM Research with contributors including the Federal Institute for Risk Assessment in Germany, the Johns Hopkins Bloomberg School of Public Health, and many others. The STEM open-source community can be found on the project website: https://www.eclipse.org/stem/ and on the wiki: wiki: http://wiki.eclipse.org/index.php/STEM. Special thanks to Dr. James Kaufman and Judith Douglas who contributed to the research and article as part of the STEM IBM project team. The development of STEM was supported in part by the Office of the U.S. Air Force Surgeon General and it was administered by the Air Force District of Washington. Nereyda Sevilla was a Biodefense PhD student in the School of Policy, Government, and International Affairs at George Mason University in Fairfax, VA. Her dissertation used STEM to examine the role of the global aviation system in relation to infectious disease outbreaks. Thanks to Dr. Gregory Koblentz, who as dissertation chair was instrumental in ensuring the completion and success of the research.

Author Contributions

The author confirms sole responsibility for the following: study conception and design, data collection, analysis and interpretation of results, and manuscript preparation.

References

1. Kirwan, D. Global Health: Current Issues, Future Trends and Foreign Policy. *Clinical Medicine*, Vol. 9, No. 3, 2009, pp. 247–253.
2. OpenFlights.org. Airport, Airline and Route Data. http://openflights.org/data.html. Accessed June 2017.
3. Sevilla, N. Open Source Disease Modeling: A Tool to Combat the Next Pandemic. *Global Biodefense*, 2016. https://globalbiodefense.com/2016/01/28/open-source-disease-modeling-a-tool-to-combat-the-next-pandemic/. Accessed June 2017.
4. National Advisory Committee. *Renewal of Public Health in Canada: Learning from SARS*. National Advisory Committee on SARS and Public Health, Ottawa, Ontario, Canada, 2003.
5. Centers for Disease Control and Prevention. H1N1 Flu. CDC Estimates of 2009 H1N1 Influenza Cases, Hospitalizations and Deaths in the United States, April–October 17, 2009. http://www.cdc.gov/h1n1flu/estimates/April_October_17.htm. Accessed June 2017.
6. U.S. Army Medical Research Institute of Infectious Diseases. *Medical Management of Biological Casualties Handbook*, 7th ed. U.S. Army Medical Research Institute of Infectious Diseases, Fort Detrick, MD, 2011.
7. Johansson, M. On the Treatment of Airline Travelers in Mathematical Models. *PLoS One*, Vol. 6, No. 7, 2011, p. e22151.
8. Lessler, J. The Cost of Simplifying Air Travel When Modeling Disease Spread. *PLoS One*, Vol. 4, No. 2, 2009, p. e4403.
9. Environmental Health & Engineering, Inc. *ACRP Report 91: Infectious Disease Mitigation in Airports and on Aircraft*. Transportation Research Board of the National Academies, Washington D.C., 2013.
10. Gaber, W. Screening for Infectious Diseases at International Airports: The Frankfurt Model. *Aviation, Space, and Environmental Medicine*, Vol. 80, No. 7, 2009, pp. 595–600.
11. Ong, R. Airline Policies and Procedures to Minimize the Spread of Disease. Presented at 89th Annual Meeting of the Transportation Research Board, Washington D.C., 2010.
12. Rydock, J. Tracer Study of Proximity and Recirculation Effects on Exposure Risk in an Airliner Cabin. *Aviation, Space, and Environmental Medicine*, Vol. 75, No. 2, 2004, pp. 168–171.
13. Mangili, A. Transmission of Infectious Diseases during Commercial Air Travel. *The Lancet*, Vol. 365, No. 9463, 2005, pp. 989–996.
14. Cooper, B. Delaying the International Spread of Pandemic Influenza. *PLoS Med*, Vol. 3, No. 6, 2006, p. e212.
15. Poletto, M. Assessing the Impact of Travel Restrictions on International Spread of the 2014 West African Ebola Epidemic. *Euro Surveill*, Vol. 19, No. 42, 2014, pp. 1–6.
16. European Centre for Disease Prevention and Control. *Infection Prevention and Control Measures for Ebola Virus Disease: Entry and Exit Body Temperature Screening Measures*. European Centre for Disease Prevention and Control, Stockholm, Sweden, 2014.
17. Wilder-Smith, A. Experience of Severe Acute Respiratory Syndrome in Singapore: Importation of Cases, and Defense Strategies at the Airport. *Journal of Travel Medicine*, Vol. 10, No. 5, 2003, pp. 259–262.
18. World Health Organization. *Effective Media Communications during Public Health Emergencies*. World Health Organization, Geneva, Switzerland, 2005.
19. Center for Disease Control and Prevention. *Crisis and Emergency Risk Communication*. U.S. Department of Health and Human Services, Washington, D.C., 2014.

The ACRP Selection Panel for the Graduate Research Award Program on Public-Sector Aviation Issues peer-reviewed this paper (18-01547).

Neither Eclipse, the U.S. Air Force, IBM, nor any of their employees, nor any contributors to STEM, make any warranty, express or implied with regard to, or assume any legal liability or responsibility for the accuracy, completeness, or usefulness of the STEM datasets. The U.S. Air Force has not accepted the recommendations made and issuance of a contract does not constitute federal endorsement of the IBM Almaden Research Center.

TRR
JOURNAL OF THE TRANSPORTATION RESEARCH BOARD

Article

Transportation Research Record
2018, Vol. 2672(29) 103–113

DOI: 10.1177/0361198118799704
journals.sagepub.com/home/trr
SAGE

Aviation Connectivity Impacts on Regional Economies in the United States

Jason C. Y. Wong[1]

Abstract

This paper uses applied microeconomics techniques to investigate the impact of aviation connectivity on 548 regional economies in the United States. Using lagged socioeconomic variables to instrument for future aviation connectivity, the paper finds a significant impact of connectivity on long-run economic growth. An increase of 100 in the city's Global Connectivity Index is associated with an increase in long-term total personal income of the city by up to $254,350,000, and up to 613 more jobs. For a city like Myrtle Beach, SC, with a connectivity index close to the mean connectivity levels of core-based statistical areas, a 100-point increase in the index represents a 1.03% increase in air connectivity. The paper also finds evidence suggesting that the impact of connectivity on regional economies is significantly more pronounced in the largest 100 cities, whereas these effects vanish in smaller cities.

Although aviation has long been regarded as the engine of globalization, there is little understanding of how aviation affects regional economic growth and development. To date, few papers have successfully shown the effect of aviation on regional outcomes; primarily because the endogeneity between air traffic activity and regional economics presents an identification challenge for econometric methods. Much like other forms of transportation, aviation provides positive externalities in enabling knowledge spillovers, such as those through networks and sociocultural exchanges and interactions. But aviation does so across unprecedented distances and speed. While it has been observed that a 5–6% growth in aviation is associated with a 2–3% annual gross domestic product (GDP) growth rate worldwide, this represents only an anecdotal relationship and correlation—it remains unclear the extent to which aviation contributes to growth, and vice versa (*1*). Estimates of the worldwide economic impact of aviation posit an effect of 3.4% of world GDP without accounting for the spillovers enabled by aviation (*1*). Thus, the casual effect of aviation on regional economies remains to be soundly identified. Building on earlier studies on investment and return on infrastructure, this research investigates how aviation contributes to local development using a new measure of aviation connectivity.

Empirical work on aviation and growth poses a significant econometric challenge because of endogeneity and limited data availability. This paper uses a novel passenger connectivity measure developed by Allroggen et al (*2*) that takes into account routing directness (including transfer time), frequency, and destination market access distance decay (*2*). To test whether there is a relationship between connectivity and growth, this paper uses lagged socioeconomic conditions to instrument for future air connectivity. The paper contributes to the empirical literature on transportation and urban growth as well as that on diffusion and endogenous growth, and in particular provides extension on the most recent work by Sheard, Blonigen and Cristea, and LeFors (*3–5*). After accounting for endogeneity, the paper finds significant impacts of a city's air connectivity on its long-run income and employment growth. This work suggests that improving the air connectivity of a city can have significant long-run impacts beyond immediate benefits. This research has the potential to inform strategic policies in sustainable aviation planning, including decisions on airport expansion, new routes, and companion investments in innovation hubs and business parks.

Literature

In the late 1980s and the early 1990s, published literature explored the value of investments in public

[1]School of International and Public Affairs, Columbia University, New York, NY

Corresponding Author:
Address correspondence to Jason C. Y. Wong: jw3144@columbia.edu

infrastructure, such as road, highways, and telecommunications. These works showed that such investments have significant impacts on economic growth (see *6*, *7*). One main approach to study empirically the effects of transport infrastructure on economic growth is to estimate the social rate of return (*6*). For example, using panel data on roadway coverage as well as regional economic data, one can estimate the returns from investing on U.S. road building. Other studies investigated how infrastructural growth in turn affected economic growth and argued for efficient transportation policy investments (*8*, *9*). Important work also investigated how railway connections improve urban economic development (*10*, *11*). Allroggen and Redding offer comprehensive reviews of the literature on transportation and regional economic development (*12*, *13*).

Aviation infrastructure differs from roads significantly. Roads provide fixed connections between places physically connected to the road, whereas aviation provides nonstop or multi-stop connections between airports depending on available routes. Operating aircraft, in comparison to personal vehicles, requires much higher entry costs and economies of scale. More importantly, air services rely on scheduling air services—although infrastructure is essential, persistent impacts of aviation anchor on the quality of connections and scheduled services that are available (*12*). Previous work studied local effects of aviation, such as on urban growth, labor market, regional output, as well as pollution and health (see *14–16*). Other attempts to establish causal links between aviation and growth or trade costs have largely rested on using air passenger or freight volume (see *13*, *17*, *18*). There has also been cross-sectional analysis using networks (*19*). Brueckner's seminal work offers a cross-section analysis using hub as an instrument and finds that expansion of airports raises service sector employment significantly (*14*).

Recent work by Blonigen and Cristea provides a quasi-natural experiment by exploiting the 1978 Airline Deregulation Act in the U.S. to study the effects of air traffic changes on local population, income, and employment. The authors find large estimated effects (*4*). Sheard uses the 1944 National Airport Plan to instrument for current airport size and distribution, with flight volume as a variable of interest. He finds positive effects of aviation on service sector employment but no effect on other sectors (*20*). In a later paper, Sheard instruments for airport size by constructing air traffic categories for aircraft, airline, and ranges, and then appending national growth onto these variables, whereby the category growth rates are orthogonal to local factors, particularly when local traffic growth is being excluded in the calculation of the national time trend (*3*). McGraw uses historical proposed air routes and mail routes to instrument for airport location and finds large effects of airports on population and employment growth (*21*). Additionally, Campante and Yanagizawa-Drott investigate air connections between city pairs and their impact on conflict (*22*); Catalini et al look at the rise of low-cost carriers and their impact on chemistry research collaborations (*23*).

Most of the literature has made use of air traffic as the independent variable of interest. LeFors constructs "air accessibility" measures, essentially a gravity measure, taking into account destination market size, as well as costs including fare and time (*5*). This research uses a time series of connectivity measures that is constructed with all scheduled passenger flights, thus building on the measure used in LeFors to better capture the potential connectedness.

Theoretical Framework

Two theoretical frameworks formally describe processes of urban economic growth and the idea exchange prompted by aviation. First, an intuitive argument can be made that the presence of aviation improves the happiness or utility of people living in the urbanity, as they have access to aviation services and the opportunity to be connected to another place with ease. For businesses, this means that firms have access to more markets, more clients, and can readily meet others in the presence of aviation connectivity. Thus, a city's income (which can be reflected through wages) and employment levels could be positively influenced by aviation connectivity. Second, aviation connectivity may be perceived instead as an enhancement in the knowledge sharing process. Having good connections to other places may mean that businesses are likely to encounter more production-improving ideas; it could also reduce the time and cost they need to interact with and process different ideas, given that good air connections exist. Therefore, aviation can induce economic growth in the city through enhancing this knowledge network. These two different, yet related, models provide the theoretical foundation of this paper. They each lead to a similar equation describing the relationship between growth and aviation connectivity.

The first framework is based on the set-up followed by Blonigen and Cristea, which augments Glaeser's model by adding air traffic growth (*4*, *24*). Essentially, aviation is seen to provide exogenous positive changes in productivity and a local amenity that factors into individuals' utility function. The simple urban growth model provides a reduced form equation for growth in population and wages with a linear relationship with aviation activity. The second framework is based on Davis and Dingel's spatial knowledge economy model that formalizes how interaction and idea exchange lead to tradables' productivity (*25*). The spatial knowledge economy model is augmented by introducing aviation connectivity in several ways. First, better aviation connectivity can increase the probability of tradables' producers encountering one another. Second, better connectivity reduces time spent on interactions, which can be allocated to produce higher value output. Finally, connectivity can be introduced as an endogenous variable determined by the economic agents.

Urban Growth Framework

This section reiterates Blonigen and Cristea closely to provide the theoretical context to compare the results of employing the connectivity measure in place of air traffic variables and to check for robustness (*4*). Their framework augments the set-up in Glaeser by introducing air services as a productivity shifter and local amenity (*24*). The same model is employed in which air traffic is replaced with the air connectivity measure for the metropolitan areas to better capture spatial spillovers.

Metropolitan statistical areas (MSAs) share common pools of labor and capital and are considered open economies. Differences in regional growth cannot therefore come from savings rates or exogenous labor supply changes. Cities differ in productivity and quality of life.

In this framework, the total output of a metropolitan area is given by the level of productivity within MSA i at time t, multiplied by the labor force raised to a national production parameter in location i at time t. Individuals earn the marginal product of labor as their wages, with normalized output prices. The model also assumes a quality of life measure that captures socioeconomic factors specific to a location that is declining in the size of the city. Individuals derive utility from their labor income and their quality of life, entering as a product.

Assuming free labor mobility across cities, each individual's utility equals the reservation utility levels at any city at any time in equilibrium. Thus, utility grows at a common rate across cities where population growth is adjusted each period to reflect changes in productivity and local amenities. This means that population and wage changes can be expressed as log-log equations with reference to changes in productivity Z_i and local amenity Q_i in the city i. By substitution, an equation relating growth of wages and population with city characteristics can be derived. Specifically, proceed by assuming a vector of city characteristics, X_{it}, for each MSA i at time t that determines the growth of both productivity Z_{it} and local amenity/quality of life Q_{it}. Moreover, the model includes a measure for air transport services as a potential growth driver for productivity and local amenity, but denoted here, c_{it}, as the aviation connectivity offered by the MSA i's airports at time t, giving:

$$\log\left(\frac{Z_{i,t+1}}{Z_{it}}\right) = (X_{it})'\gamma_1 + \beta_1 \log\left(\frac{c_{i,t+1}}{c_{it}}\right) + \nu_{it}$$

$$\log\left(\frac{Q_{i,t+1}}{Q_{it}}\right) = (X_{it})'\gamma_2 + \beta_2 \log\left(\frac{c_{i,t+1}}{c_{it}}\right) + \upsilon_{it}$$

By substitution the reduced form for log growth rates may be derived as:

$$\dot{L}_i = \tilde{\beta}_1 \dot{c}_i + X'_{i,T_0}\tilde{\gamma}_1 + \varepsilon_{it} \quad (1)$$

$$\dot{W}_i = \tilde{\beta}_2 \dot{c}_i + X'_{i,T_0}\tilde{\gamma}_2 + \xi_{it} \quad (2)$$

with $\tilde{\gamma}_1, \tilde{\gamma}_2, \tilde{\beta}_1, \tilde{\beta}_2$ to be estimated empirically. The key hypothesis is that improved connectivity will yield increased population growth, per capita and regional income levels, employment, and other measures of changes in regional economic activity, subject to time-invariant city characteristics.

Connectivity in a Spatial Knowledge Economy

This section adapts an idea-exchange model developed by Davis and Dingel and augments it by introducing aviation connectivity (*25*). The model makes explicit that knowledge gained in interacting with others, in exchanging ideas, is not costless. There is a time-opportunity cost involved. The model also assumes a tacit nature of knowledge, requiring physical interaction or face-to-face communication and, as such, cities are the loci of these interactions. Finally, labor is heterogeneous and continuously distributed.

The spatial knowledge economy involves individuals consuming three goods: tradables, non-tradable services, and non-tradable housing. The indirect utility function is given by:

$$V(p, I) = I_c - p_{s,c}\bar{s} - p_{h,c}$$

in which consumers have income I, face prices p in city c, expend on $\bar{s}$, the non-tradables, and the remaining income on tradables (numeraire). Consumers are perfectly mobile across sectors and across cities; their choices of location and occupation maximize $V(p, I)$.

Assuming non-tradables are produced at uniform levels of productivity, for a mass of worker L with heterogeneous abilities $z \backsim \mu(z)$, low ability people in this model will specialize in non tradables by comparative advantage. In the tradable sector, producers can gain knowledge to increase productivity, and individual productivity in tradables $\tilde{z}(z, Zc)$ is thus increasing in z and depends on the set of learning opportunities available in the city, Z_c, such as availability to interact with others.

Tradable producers have one unit of time to divide between production (β units of time) and interaction ($1 - \beta$ units of time). An agent with ability z produces tradables output:

$$\tilde{z}(z, Z_c) = \max_{\beta \in [0,1]} \beta z(1 + (1-\beta)GZ_c z)$$

G denotes the productivity benefits of learning, which is common to all cities, to reflect gains from interactions. When G is higher, interactions are more beneficial for productivity. or interactions benefit productivity more.

The opportunity to learn and interact within a city, Z_c, is determined by random matching processes in which producers take time to encounter and exchange ideas, and the original model can be augmented with the connectivity of the city to other places (thus have higher exposure to possible ideas). Consequently, Z_c is defined as follows:

$$Z_c \equiv m(M_c, A_c)\bar{z}_c$$

where

m = the probability of encounter and an increasing function of its two arguments,

M_c = the total time spent by producers in interacting,

A_c = the aviation connectivity measure for the city,

$\bar{z}_c$ = the city's expected ability levels.

With this augmentation, A_c is treated as an exogenous feature of the city c not determined by consumers or producers in this model. One can proceed to solve for the equilibrium in the set of cities with population L for the equilibrium prices, locations, time spent on sectors, occupations.

In the above augmentation, aviation is treated as having no effect on the choice of time to interact: aviation simply improved the probability of encounters. However, one can expect that improved connectivity also reduces the time that producers need to interact for idea exchanges. So the next modification further introduces aviation as a factor that reduces the total time spent interacting, that is:

$$M_c = f(L, \beta, \mu(z, c), A_c)$$

This means that more time can be spent on production, with higher connectivity in addition to aviation providing more amenity in city c. This leads to higher output levels.

Finally, an ideal augmentation would introduce aviation as an endogenous variable by adding air services as a sector. This is likely to complicate the model and be difficult to solve, but essentially the air sector produces products that add to local amenity as well as reduces tradables' production and interaction costs. Thus, the economy will allocate efficiently the ideal proportions of labor to each sector to maximize output. Multiple equilibria are expected.

With the equilibria in mind, linear model of Z_c is assumed as follows:

$$Z_c = \gamma_0 + \gamma_1 A_c + \gamma_2 M_c(\bar{z}_c) + X_c'\vec{\gamma} + \varepsilon_c \quad (3)$$

where

A_c = the aviation connectivity measure for the city,

$M_c(\bar{z}_c)$ = the total time spent by producers in interacting, which is an increasing function of the city's expected ability,

X_c = the vector of city characteristics that determines whether a positive learning environment is provided,

ε_c = the error that also captures the stochastic idea matching process.

Holding all else constant, the equilibrium conditions include aviation connectivity, where increases in A_c should, ceteris paribus, lead to increases in production of tradables.

To sum up, the two models presented above yield three key equations (Equations 1–3). These equations all describe how regional economic outcomes on the left hand-side can be described as a linear function of aviation connectivity (c in the first model, in Equations 1–2; A in the second model, in Equation 3) and other explanatory variables based on the model, including city characteristics and other factors. In the following section on the econometric model, the empirical model will be based on these equations with minor modifications or addition of other explanatory variables; c will be used as the notation for aviation connectivity, consistent with the first model above.

Econometric Model

Identification Challenges

The key challenge to estimate the effects of aviation and local development empirically is endogeneity. Clearly, economic growth causes investment in airports, route licensure, and bid for take-off/landing slots. Such simultaneity may bias the estimates for marginal impacts of aviation on the economy. Using instrumental variables would be one valid approach in lieu of taking increased aviation activity as exogenous. Various studies have also included country and year fixed effects to de-trend macroeconomic environments at play.

The aforementioned studies use an outcome variable—either flight traffic, passenger, or freight volume (*3*, *4*, *20*, *21*), or in rarer cases accessibility (*5*), as well as instrumental variables such as hub status or an older policy—and run a 2SLS to estimate effects of airport size or air traffic on local development.

Following the theoretical framework, the equation of interest is:

$$y_i = \alpha + \beta c_i + x_{2,i}'\gamma + \varepsilon_i \quad (4)$$

where

y = the outcome variable of interest in region i, such as output and employment,

c = the connectivity measure,

$x_{2,i}$ = the set of control variables,
ε = the errors.

As c is likely endogenous, 2SLS is performed using an instrument for connectivity. First, connectivity is regressed on the instrumental variable z and some other controls:

$$c_i = \phi + \eta z_i + x'_{1,i}\delta + u_i \quad (5)$$

and predicted connectivity values $\hat{c}$ are obtained. Then $\hat{c}$ is used in Equation 1 in which outcome variables y are regressed on the predicted connectivity values $\hat{c}$. Identification thus requires two conditions—the relevance and exogeneity conditions—one being that the instrument has to be correlated with the independent variable, that is, $\eta = 0$, and the other being that the instrument must be exogenous to the outcome.

Given the initial data availability, the basic model of interest in this paper will take the following form:

$$\Delta y_{i,2012-1990} = \alpha + \beta c_{i,2012} + x'_{2,i}\gamma + \varepsilon_i \quad (6)$$

Essentially, this is a cross-section analysis for the 30-year change in income and employment or other socioeconomic outcome variables (between 2012 and 1990), with the connectivity measure recorded at year 2012. c is then instrumented by lagged socioeconomic variables (base year of 1970).

Data

Independent Variable

Connectivity networks are increasingly considered as important frameworks in understanding various economic behaviors (*19*, *26*). Allroggen et al developed a new metric for measuring connectivity using route schedules and hub centrality network measures, available from 1990 to 2012 (*2*). Compared to previous constructed measures, the Global Connectivity Index (GCI) by Allroggen et al improves on the methodology by incorporating network theory, gravity of destination quality, directness of routes, and sensible opportunity cost.

The GCI is constructed as follows:

$$GCI_{a,t} = \sum_{r \in R_{a,t}} \alpha_{a,t} f_{a,t} w_{d_r,t} \quad (7)$$

where

a = airport,
d = the destination airports,
$r \in R$ = a route within the set of all routings,
α = directness,
f = the frequency,
w = the destination quality.

This new measure offers a better view of the network access of aviation compared with just passenger volume, for example, which gives measurement error by disregarding connections, directness, layover times as opportunity costs, and destination access. The GCI is then converted into the aggregated core-based statistical areas (CBSA) level connectivity measure c_i by geospatially locating the airports into the CBSA. Figure 1 shows the airport–CBSA correspondence for parts of the United States. The shaded polygons are the shapes of each CBSA in the United States, and the green icons indicate the airports that are located within a CBSA. Note that it is possible for one CBSA to contain more than one airport. In such cases, GCI scores are summed linearly within these CBSAs. Of the 929 CBSAs in the United States, there are 1,039 airports that fall within 548 CBSAs, as there are several CBSAs that contain no airports in their defined geography.

Dependent Variables

The dependent variables of interest include socioeconomic variables such as income and employment levels. At the county and MSA level, business and socioeconomic data are readily available from the U.S. Census Bureau and the Bureau of Economic Analysis. These variables include population, personal and per capita income, and employment for all CBSAs from 1970 to 2012. The data are then combined with the geospatially integrated connectivity data at the CBSA level.

Instrumental Variable

Whereas several studies in the existing literature have constructed instruments based on historical plans or policies, fraction of connecting passengers, concentration of passengers originating from urban areas, or national trends (*3–5*, *20*), this paper uses historical socioeconomic data to instrument for future connectivity. A valid instrument for the GCI in the present (in 2012) would need to correlate with connectivity today but not with the outcome variable of interest. It can be argued that income, population, and per capita income in 1970 correlate with future connectivity, as past income and population are necessary conditions for connectivity in the future through flight routes and connections. However, levels of past income and population need not necessarily affect future change or growth in income or in innovation. Although recognizing that this may not be the best instrument, the paper uses lagged socioeconomic variables as instruments for connectivity to gauge a first-cut effect of aviation connectivity on long-run economic changes at the city level and to see if consistent results with the literature can be documented.

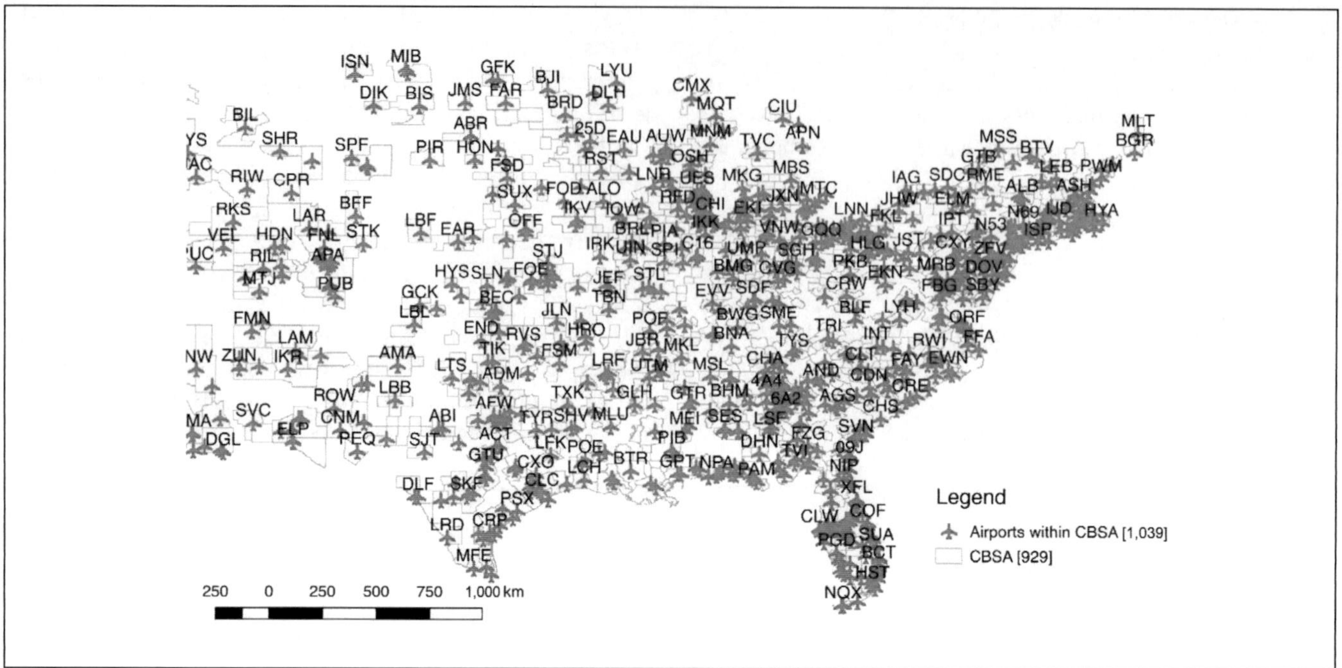

Figure 1. Matching airport-level data onto CBSAs in the United States.

Results

The following results tables report the 2SLS regression results for different sets of outcome and control variables. The key regression equation of interest is the regression of the long-term change in socioeconomic outcomes (1990–2012) on aviation connectivity levels in 2012 for 548 airport- containing CBSAs. In Tables 1–4, the odd columns report the first-stage results and the Sanderson-Windmeijer (SW) *F* statistics to show the validity of the chosen instruments, generally taken as an *F*-stat > 10. In Columns 1–2, the instrument for the sum of global connectivity in 2012 is the income in 1970. It is assumed that the income levels in 1970 are correlated with future air connectivity through aviation infrastructural investments, but do not have an impact on long-term changes in the outcome. In Columns 3–4, GCI 2012 is instrumented for with population and per capita income in 1970 instead.

Table 1 considers the outcome variable of income change between 1990 and 2012. The instruments are demonstrated to be valid in the first stage. A significant impact of connectivity robust to both specifications of instruments is found. Take the example of the first model (results displayed in Columns 1–2), which uses the 1970 income to instrument for 2012's level of connectivity. The coefficient of 2,543.5 is positive and significant, meaning that the higher the connectivity levels in 2012, the higher the growth in income between 1990 and 2012. Interpreting this strictly in the linear sense, a city that has a 100-unit higher level of the GCI can be associated with a higher level of long-term total personal income by up to $254,350,000. Somewhat surprisingly, the change in employment is negatively related to change in personal income, but this is not significant. In the second model (results displayed in Columns 3–4), which uses the alternate set of lagged socioeconomic variables as instruments, the results are found to be qualitatively similar, that is, the magnitudes and signs of the coefficients are similar (slightly smaller).

Table 2 considers the change in per capita income between 1990 and 2012. In this specification, significant impacts of connectivity to per capita income growth are found at the 0.05 level and 0.01 level respectively for the two sets of instruments. In the first two columns, a positive impact of connectivity on per capita income growth by 0.06 is found at the 0.05 level; employment change's relationship to per capita income growth remains insignificant. In Columns 3-4, in which the alternate set of instruments for connectivity is applied, the magnitude of the coefficient is twice as large as the previous model in Column 2. For a 100-point increase in the GCI levels in 2012, growth of long-run per capita income increases by $10.60, according to the coefficient in Column 4. The R^2 value suggests, however, that unlike personal income, a linear model is less suitable to explain the variation in the city per capita income changes. Overall, the model specified in Table 2 does not provide strong evidence on how connectivity affects per capita income. Taking both the stronger results in Table 1 and the weaker results in Table 2 together, a potential omitted factor could be significant population changes in areas with higher aviation activity.

Table 1. 2SLS Results: Δ Income = Change in Total Personal Income between 1990 and 2012

	(1) 2012 GCI sum	(2) Δ Income	(3) 2012 GCI sum	(4) Δ Income
Income in 1970	0.00220***			
	(0.000604)			
Δ Employment	0.0788***	−91.66	0.0751***	−75.51
	(0.00887)	(65.71)	(0.00872)	(57.02)
2012 GCI sum		2,543.5***		2,409.8***
		(611.6)		(544.8)
Population in 1970			0.0113***	
			(0.00272)	
Per capita income in 1970			2.356**	
			(0.847)	
SW F statistic	13.31		20.62	
SW p-value	0.000289		2.33e–09	
N	548	548	548	548
Partial R^2	0.472		0.472	
R^2		0.790		0.810

Note: Standard errors in parentheses. Odd columns show first-stage results with SW first-stage F statistics. Even columns show second-stage results.
$^{\dagger}p < 0.10$, $^{*}p < 0.05$, $^{**}p < 0.01$, $^{***}p < 0.001$.

Table 2. 2SLS Results: Δ Per Capita Income = Change in Per Capita Income between 1990 and 2012

	(1) 2012 GCI Sum	(2) Δ Per capita income	(3) 2012 GCI sum	(4) Δ Per capita income
Income in 1970	0.00220***			
	(0.000604)			
Δ Employment	0.0788***	−0.00112	0.0751***	−0.00678
	(0.00887)	(0.00344)	(0.00872)	(0.00438)
2012 GCI sum		0.0586*		0.106**
		(0.0275)		(0.0368)
Population in 1970			0.0113***	
			(0.00272)	
Per capita income in 1970			2.356**	
			(0.847)	
SW F statistic	13.31		20.62	
SW p-value	0.000289		2.33e–09	
N	548	548	548	548
Partial R^2	0.472		0.472	
R^2		0.0260		0.0230

Note: Standard errors in parentheses. Odd columns show first-stage results with SW first-stage F statistics. Even columns show second-stage results.
$^{\dagger}p < 0.10$, $^{*}p < 0.05$, $^{**}p < 0.01$, $^{***}p < 0.001$.

Table 3 reports the results with the change in employment levels as the outcome variable, controlling for population growth. The GCI again has significant impacts to long-term employment changes in both sets of instruments. In the first two columns, positive and significant impacts of both connectivity and population growth on long-run employment changes are found. Applying the coefficient for connectivity directly in Column 2, long-run employment growth of up to 234.6 jobs can be associated with a 100-point higher levels of connectivity. The alternative set of instruments yields qualitatively similar results, albeit with a slightly smaller magnitude effect from connectivity and slightly larger population effects.

Table 4 also reports results with employment growth but controlling for per capita income instead of population growth. The results are also significant. In the first two columns using the first set of instruments, a positive and significant coefficient of 6.096 is found; using the second set of instruments and applying the coefficient, a 100-point increase in the GCI for a city can be associated with 613 more jobs. Whereas explanatory power decreases in these two specifications, the magnitude of

Table 3. 2SLS Results: Δ Employment = Change in Total Employment between 1990 and 2012

	(1) 2012 GCI sum	(2) Δ Per capita income	(3) 2012 GCI sum	(4) Δ Per capita income
Income in 1970	0.00266***			
	(0.000498)			
Δ Population	0.0439***	0.442***	0.0418***	0.446***
	(0.00513)	(0.0532)	(0.00503)	(0.0496)
2012 GCI sum		2.346**		2.290**
		(0.868)		(0.814)
Population in 1970			0.0136***	
			(0.00216)	
Per capita income in 1970			2.180**	
			(0.776)	
SW *F* statistic	28.49		34.54	
SW *p*-value	0.000000138		7.55e–15	
N	548	548	548	548
Partial R^2	0.616		0.616	
R^2		0.969		0.969

Note: Standard errors in parentheses. Odd columns show first-stage results with SW first-stage *F* statistics. Even columns show second-stage results.
$^{\dagger}p < 0.10$, $^{*}p < 0.05$, $^{**}p < 0.01$, $^{***}p < 0.001$.

the effect is more than doubled from Table 3. The results in Table 4, considering the magnitude and explanatory power differences, can be interpreted as an upper-bound effect.

Finally, Table 5 reports a series of specifications for large versus small cities. In these specifications, the GCI is instrumented for by both population and per capita income in 1970. In Columns 1–2, connectivity positively affects cities—and the magnitude of the impact in the top 100 cities is more than twice that in the bottom 267 CBSAs. Moreover, the R^2 is indicative that the smaller cities cannot be explained by the linear model imposed. For employment in Columns 3–4, the impact of connectivity is positive and significant for the largest CBSAs; the effect vanishes for the smaller CBSAs, and again, the linear model fails to account for the variation in the data. Smaller cities seem to benefit less or insignificantly from connectivity boosts; further work should investigate potential non-linearities.

Discussion and Conclusion

Few studies have investigated the impact of aviation connectivity on long-run economic growth and innovation. Aviation activity itself is endogenous with local economic conditions and this presents a challenge for empirical identification of the causal impacts. As a first cut to verify whether connectivity is robust to this kind of analysis, this paper uses lagged socioeconomic variables to instrument for future air connectivity to study the impact of air connectivity on cities' economies. Instead of air passenger numbers, which are typically used in past literature to measure aviation activity, this study uses the GCI, which better captures the access aviation provides to a city through the network and alliances it uniquely embodies with codeshares and hub structures. In a rapidly globalizing world, the connectivity provided by aviation could have unprecedented spillovers into enabling the development of new ideas and capital.

Using an instrumental variable approach, this research finds a significant impact of connectivity to long-run economic growth. For example, a city like Myrtle Beach, SC, has a connectivity index close to the mean connectivity levels of CBSAs. Given the GCI for Myrtle Beach, a 100-point increase in the connectivity index would represent or require a 1.03% increase in air connectivity. For reference, the total employment level in 2012 for Myrtle Beach was 189,007 and total personal income was $12,517,846,000. Applying the marginal effects computed above, the hypothesized 100-point increase (1.03%) represents a marginal effect of 2.03% in total income and 0.3% in employment for the mean connectivity city.

This paper provides some evidence to support policies that aim to increase aviation connectivity in different regions. The use of air connectivity in this research implies that cities and regional governments could aim beyond merely increasing passenger numbers or the number of flights. Adding new connections to large hubs can allow the city to enjoy large connectivity benefits; passengers thereby benefit from more one-stop flights, lower layover times, and codesharing opportunities. Adding new direct flight destinations can also improve nonstop connectivity, bringing about connectivity-driven spillover impacts to a region. Regional governments can also support connectivity-based growth by investing in airport infrastructure (both on-airport and companion

Table 4. 2SLS Results: Δ Employment = Change in Total Employment between 1990 and 2012

	(1) 2012 GCI sum	(2) Δ Per capita income	(3) 2012 GCI sum	(4) Δ Per capita income
Income in 1970	0.00420***			
	(0.000677)			
Δ Per capita income	0.167†	−0.00236	0.0710	−0.0211
	(0.0880)	(0.410)	(0.0692)	(0.407)
2012 GCI sum		6.096***		6.130***
		(0.546)		(0.533)
Population in 1970			0.0210***	
			(0.00295)	
Per capita income in 1970			3.746**	
			(1.261)	
SW *F* statistic	38.49		50.59	
SW *p*-value	1.09e–09		7.09e–21	
N	548	548	548	548
Partial R^2	0.713		0.713	
R^2		0.787		0.787

Note: Standard errors in parentheses. Odd columns show first-stage results with SW first-stage *F* statistics. Even columns show second-stage.
†$p < 0.10$, *$p < 0.05$, **$p < 0.01$, ***$p < 0.001$.

Table 5. 2SLS Results: Large versus Small Cities

	(1) Δ Income	(2) Δ Income	(3) Δ Employment	(4) Δ Employment
2012 GCI sum	2,459.6***	1,147.6***	5.808***	−0.227
	(462.0)	(250.8)	(0.763)	(1.603)
Δ Employment	−55.87	−4.721		
	(43.80)	(25.78)		
Δ Per capita income			−1.105	0.269**
			(3.495)	(0.0873)
N	100	267	100	267
R^2	0.797	−2.292	0.698	0.0678

Note: Standard errors in parentheses. Columns 1 and 3 are limited to the Largest 100 CBSAs; Columns 2 and 4 are limited to the Smallest 267 CBSAs.
†$p < 0.10$, *$p < 0.05$, **$p < 0.01$, ***$p < 0.001$.

transportation networks), zoning and development of business or tech hubs, and other companion investments. These actions would lower entry costs and attract air services. In addition to market demand studies, stakeholders should take into account the connectivity-driven impacts when considering actions to expand air services.

Of the CBSAs studied, this research also finds evidence to support a "multiple equilibria" theory in the urban economics literature. Large cities have higher connectivity benefits: the magnitude and significance of connectivity impacts to the top 100 CBSAs in the United States are pronounced; these effects vanish when the smallest 30% of the CBSAs with airports are studied. This provides evidence to the following: (1) there may be non-linear impacts of connectivity in small cities; and (2) there may be threshold effects of connectivity—a city will need to acquire a certain level of base connectivity beyond which connectivity has positive externalities and persistent impacts in the local economy, akin to a "low connectivity poverty trap."

Further Work

This research contributes to the existing literature by applying the latest network connectivity measure in the transportation literature to the literature in economic geography. The study uses an empirical set-up to estimate the effects of aviation activity on local economies. Further work is needed to refine the theoretical framework as well as the identification strategy. Testing the robustness of results using new instruments would be ideal; the lagged variable approach in place may not fully satisfy exclusion restrictions, but nonetheless provides good insight into an upper-bound effect of aviation on regional economics. Additional outcome variables such as patents that directly measure innovation should be explored.

Noting that the linear model does not fit the variation of small MSA data well, non-linear methods for the lower quantiles of the cities should be explored. In addition, Arellano-Bond linear dynamic panel-data estimation would offer a method to test the full data set; currently, the paper uses a subset of what is available in the master panel for simplicity and to outline whether key relationships can be tested. Finally, a potential extension is to investigate the added value of aviation to cities through hedonic approaches, but the data requirement on residential housing prices and local airport accessibility may be cumbersome.

Acknowledgments

I thank the Transportation Research Board Airport Cooperative Research Program Graduate Research Award (ACRP 11-04) for the financial support and mentorship for this work. I especially want to thank Dr. Florian Allroggen (MIT) and Professor Dr. Robert Malina (Hasselt) for their mentorship and for providing data and motivation for this work. I want to thank Professor Steven Barrett and his lab, Laboratory for Aviation and the Environment, for hosting me during the Fall 2016 semester and for having me as a Research Affiliate. I am also grateful to all seminar/colloquium participants for helpful feedback at Columbia (Sustainable Development; Urban Economics), MIT (Laboratory for Aviation and the Environment), and Rutgers (Krueckeberg Conference in Urban Studies).

References

1. Belobaba, P. Introduction and Overview. In *The Global Airline Industry*, 2nd ed. (Belobaba, P., A. Odoni, and C. Barnhart, eds.), John Wiley & Sons, Chichester, UK, 2016, pp. 1–18.
2. Allroggen, F., M. D. Wittman, and R. Malina. How Air Transport Connects the World—A New Metric of Air Connectivity and Its Evolution Between 1990 and 2012. *Transportation Research Part E: Logistics and Transportation Review*, Vol. 80, 2015, pp. 184–201.
3. Sheard, N. *Airport Size and Urban Growth.* Technical Report, Economica: https://onlinelibrary.wiley.com/doi/full/10.1111/ecca.12262
4. Blonigen, B. A., and A. D. Cristea. Air Service and Urban Growth: Evidence from a Quasi-natural Policy Experiment. *Journal of Urban Economics*, Vol. 86, 2015, pp. 128–146. https://doi.org/10.1016/j.jue.2015.02.001.
5. LeFors, M. Essays on Transportation Innovations and Local Development. Order No. 3734468, University of California, Irvine. https://search.proquest.com/docview/1748046287?accountid = 10226. 2015.
6. Canning, D., and M. Fay. *The Effect of Transportation Networks on Economic Growth.* Department of Economics Discussion Paper 9293, Columbia University, 1993.
7. Aschauer, D. A. Is Public Expenditure Productive? *Journal of Monetary Economics*, Vol. 23, No. 2, 1989, pp. 177–200. https://doi.org/10.1016/0304-3932(89)90047-0.
8. Wylie, P. J. Infrastructure and Canadian economic growth 1946–1991. *Canadian Journal of Economics*, Vol . 29, No. 1, 1996, pp. 350–355.
9. Winston, C. Efficient Transportation Infrastructure Policy. *Journal of Economic Perspectives*, Vol. 5, No. 1, 1991, pp. 113–127. https://doi.org/10.1257/jep.5.1.113.
10. Haines, M. R., and R. A. Margo. *Railroads and Local Economic Development: The United States in the 1850s.* NBER Working Paper Series 12381, National Bureau of Economic Research, Cambridge, MA, 2006.
11. Donaldson, D., and R. Hornbeck. Railroads and American Economic Growth: A "Market Access" Approach. *The Quarterly Journal of Economics*, Vol. 131, No. 2, 2016, pp. 799–858. https://doi.org/10.1093/qje/qjw002.
12. Allroggen, F. Fördert ein leistungsfähiges Verkehrssystem die wirtschaftliche Entwicklung? Ein Überblick. *Zeitschrift für Verkehrswissenschaft*, Vol. 84, No. 3, 2013, pp. 2–43.
13. Redding, S. J., and A. M. Turner. Transportation Costs and the Spatial Organization of Economic Activity. *Handbook of Urban and Regional Economics*, Vol. 5, 2015, pp. 1339–1398. https://doi.org/10.1016/B978-0-444-59531-7.00020-X.
14. Brueckner, J. Airline Traffic and Urban Economic Development. *Urban Studies*, Vol. 40, No. 8, 2003, pp. 1455–1469. https://doi.org/10.1080/0042098032000094388.
15. Allroggen, F., and R. Malina. Do the Regional Growth Effects of Air Transport Differ among Airports? *Journal of Air Transport Management*, Vol. 37, 2014, pp. 1–4. https://doi.org/10.1016/j.jairtraman.2013.11.007.
16. Wolfram, S., and W. R. Walker. Airports, Air Pollution, and Contemporaneous Health. *Review of Economic Studies*, Vol. 83, No. 2, 2016, pp. 768–809. https://doi.org/10.1093/restud/rdv043.
17. Anderson, J. E., and E. V. Wincoop. Trade Costs. *Journal of Economic Literature*, Vol. 42, No. 3, 2004, pp. 691–751.
18. Button, K., and J. Y. Yuan. Airfreight Transport and Economic Development: An Examination of Causality. *Urban Studies*, Vol. 50, No. 2, 2013, pp. 329–340. https://doi.org/10.1177/0042098012446999.
19. Guimerà, R., S. M. A. Turtschi, and L. A. N. Amaral. The Worldwide Air Transportation Network: Anomalous Centrality, Community Structure, and Cities' Global Roles. *Proceedings of the National Academy of Sciences of the United States of America*, Vol. 102, No. 22, 2005, pp. 7794–7799. https://doi.org/10.1073/pnas.0407994102.
20. Sheard, N. Airports and Urban Sectoral Employment. *Journal of Urban Economics*, Vol. 80, 2014, pp. 133–152. https://doi.org/10.1016/j.jue.2014.01.002.
21. McGraw, M. J. *Perhaps the Sky's the Limit? Airports and Employment in Local Economies.* UC Berkeley, 2015.
22. Campante, F., and D. Yanagizawa-Drott. *Long-Range Growth: Economic Development in the Global Network of Air Links.* UBS Center Working Paper No. 16, UBS International Center of Economics in Society, Zurich, 2016.
23. Catalini, C., C. Fons-Rosen, and P. Gaull. Did Cheaper Flights Change the Direction of Science? *SSRN Electronic*

Journal, Vol. 9897, 2016, pp. 1–27. https://doi.org/10.2139/ssrn.2764219.

24. Glaeser, E. L., J. A. Scheinkman, and A. Shleifer. Economic Growth in a Cross-section of Cities. *Journal of Monetary Economics*, Vol. 36, No. 1, 1995, pp. 117–143. https://doi.org/10.1016/0304-3932(95)01206-2.
25. Davis, D. R., and J. I. Dingel. *A Spatial Knowledge Economy*. NBER Working Paper 18188. National Bureau of Economic Research, Cambridge, 2012.
26. Jackson, M. O. Networks in the Understanding of Economic Behaviors. *The Journal of Economic Perspectives*, Vol. 28, No. 4, 2014, pp. 3–22. https://doi.org/10.1257/jep.28.4.3.

The ACRP Selection Panel for the Graduate Research Award Program on Public-Sector Aviation Issues peer-reviewed this paper (18-04034).

Sage Publications USA
sagepub.com

ISBN 978-1-5108-9205-7

CURRAN ASSOCIATES INC.
proceedings.com